9
2

The Genesis and Evolution of Time

The Genesis
and Evolution of Time

A Critique of Interpretation in Physics

J. T. Fraser

The University of Massachusetts Press
Amherst, 1982

Beneath this new and open landscape
the steady devotion of my wife

Acknowledgments

I am deeply indebted to Professor Nathaniel Lawrence for his careful, critical reading of the manuscript. His scholarship enriched both the book and the mind of its author.

I wish to thank the Library of Manhattanville College for the stack privileges I have enjoyed for well over a decade, the Burndy Library for its ever ready welcome and the Westport Public Library for its reliable interlibrary service.

A workshop held at the invitation of the Rockefeller Foundation at their Bellagio Study and Conference Center helped clarify some of the issues discussed in the book.

Illustration credits are shown on the appropriate pages.

Contents

The Genesis and Evolution of Time

Introductory Remarks

Let us assume that time is a symptom or correlate of the structural and functional complexity of matter. It is a generally accepted hypothesis of modern science that the dynamics of the universe is one of inorganic and organic evolution. It would follow that time itself has evolved with the increasing complexification of natural systems.

It is easy to demonstrate that the concept of time has undergone many historical changes, but this is not the claim. The proposition is that time had its genesis in the early universe, has been evolving, and remains developmentally open-ended.

The notion of time as having a natural history is difficult to assimilate within received teachings or even to express in noncontradictory statements. Yet the detailed inquiry carried out in this monograph reveals that the evolutionary character of time is already implicit in the ways time enters physical science in particular and natural science in general.

The interpretative proposition made and examined in this book is called the principle of temporal levels. It maintains that each stable integrative level of the universe manifests a distinct temporality and that these temporalities coexist in a hierarchically nested, dynamic unity.

The principle of temporal levels offers substantial economy of thought for dealing with time in special relativity theory, quantum theory, thermodynamics, general relativity theory, and the thermodynamics of biogenesis. It elucidates a number of empirical and theoretical issues for which, thus far, only ad hoc explanations have been available, while also revealing an unsuspected unity among the major theorems of physics. Furthermore, it permits the tracing of a continuous path connecting the time-related findings of physics, biology, psychology, and sociology. A coherent unity of the different domains of modern science is thus suggested, without

challenge to the necessarily distinct methods of specialized knowledge.

This book employs very little mathematics. However, familiarity with the physicist's way of thinking and with the major issues of
modern physics is assumed. Appeal is frequently made to the simple
experimental facts and conceptual arguments upon which the
sophisticated theorems of modern physics are based. This is never
done, however, to introduce those theorems for the first time, but
rather to refresh the reader's memory and to enable us to see the
theoretical and empirical issues involved in a new perspective,
afforded by the principle of temporal levels.

1

The Measurement
of Time

What do we do when we measure time? This chapter proposes to show that each measurement of time is an experiment that tests a hypothesis about nature. It concludes that if we wish to learn about time, clock readings in themselves will be of little use. We must consult, instead, the hypotheses connecting those readings.

1.1 Keeping Time

Imagine a fencepost casting its shadow upon flat ground. Observe the changing directions and lengths of the shadow as the sun moves from morning till night and as its orbit changes with the seasons. Construct a polar coordinate system of directions and circular arcs to measure the changing directions and lengths of the shadow. Use such angular divisions and radial lengths as appear practical. Identify the lines and arcs of the geometrical design with consecutive, whole real numbers: so many different directions, so many different lengths. You have constructed a clock of great universality and sophistication.

The fencepost became the gnomon of the sundial, taking its name from the Greek γνωμων meaning adviser, inspector, or one who knows. The name is appropriate because the gnomon serves as one who knows, one who interprets. Just as an interpreter provides continuous translation of one language into another, thereby joining two linguistic universes according to certain principles of transformation, so the gnomon joins two processes according to certain hypotheses. One of the processes is the motion of the sun; the other is the life of the dialer; the hypothesis comprises the principles implicit in the construction and operation of the sundial.

The sundial is part of the universe of man: it is a human artifact whose mission is to regulate the life of its maker. It is also a part of the astronomical universe: the motion of the shadow is independent of human volition. It recognizes and unites these two worlds because of the conviction of the dial maker that it is possible to derive lawful time signals from the motion of the shadow. This belief permits him to arrange his life according to the motion of the shadow and prohibits him from expecting the shadow to progress according to the rhythm and tempo of his own needs and desires.

Imagine now the life of the dialer. One sunny day when the shadow points to numeral *14,* a friend appears with a wheel of cheese and when the shadow points to numeral *8,* a dog barks. These conditions amount to the recognition of two sets of simultaneities. "Shadow *14* + friend with cheese" is one, "shadow *8* + bark of dog" is the other. The two sets of simultaneities, remembered ex post facto, allow the dial watcher to declare that the friend-with-cheese event was separated from the bark-of-dog event by *6* or *18* arbitrary units of time, provided he divides his days into *24* hours and depending on which of the two events happened earlier, as judged by his memory.

This method of keeping time is validated by a number of hidden assumptions. For instance, the dial maker took it for granted that the progress of the sun along its orbit from shadow *8* to *14* or *14* to *8,* and the progress of his life from cheese to bark or from bark to cheese, are two continuous processes whose consecutive states may be meaningfully correlated, instant by instant. He need not know any specific laws that govern the correlation but he must have some practical knowledge corresponding to the idea of lawful interdependence.

The numerals identifying the different directions and lengths of the shadow may be imagined as values of the independent variable of a transformation equation, one that changes the numbers assigned to shadow-instants to the numbers assigned to life-instants, and vice versa. It would be immensely difficult to write such a transformation equation if even one of the processes were as complex as the one involving the man, the friend, the cheese, and the dog. But it is not so difficult to do so if the processes are less complex, such as, for instance, the spinning of the earth about its axis on the one hand and, on the other hand, the oscillations of atoms in a magnetic field. All equations of physics belong in this second, relatively simple category of transformations.

Using the sundial and its bucolic setting as a paradigm, we are ready to work out a vocabulary for time measurement.

Any happening singled out for attention is an *instant*. Each distinct direction or length of the shadow defines an instant of time as does the coming of the friend. Two locally simultaneous instants make up an *event*. "Friend at shadow *14*" is an event, "bark at shadow *8*" is another one. If the dog is far away then the "hearing of the bark at shadow *8*" is the event.

The temporal separation between two events is just that: *temporal separation* or *time passed*. If the earlier event happened very long ago and is of special importance, then the temporal separation is the *epoch* of the later event. But the usage is quite loose. *Reference epoch* sometimes means the earlier event with respect to a yet earlier one—or a later event.

With this vocabulary to be employed as needed, we turn to selected representative examples of time keeping. They are classed under a number of useful but not mutually exclusive headings. Those familiar with the rich history of timekeepers will find the presentation a rough ride, especially because we must jump back and forth across centuries or even millennia. But the show has a purpose: it will help us formulate certain general rules for the measurement of time.

Celestial cycles The earliest known devices using the diurnal rotation of the sun for the measurement of time are Egyptian shadow clocks dating from the eighth century B.C. According to literary and archeological evidence, sundials first appeared in the sixth century B.C. The earliest preserved examples, however, are of later vintage. Figure 1 shows a sundial believed to be from the third century B.C. It was found in 1873 at Heraclea at Latmum, an ancient town in western Asia Minor, in what is now Turkey.

The gridwork engraved upon its conical surface suggests that the dial was used as a daily clock (the direction of the shadow indicating the hour) as well as a yearly clock, better known as a calendar (the length of the shadow at noon indicating the season). The identification of the straight hour lines and the day curves by different numbers and names constituted a crucial step in the art of time measurement. Namely, it reduced the need for specifying the direction or size of the shadow with respect to other objects to the naming of names and numbers referring to a structure (the sundial) whose position was determined by convention.

Figure 1. Front view of a conical sundial found in 1873 at the foot of Mount Latmus, Turkey. The face of the sundial is a cone whose axis is placed in parallel with the axis of the earth. The front of this white marble dial, the plane with the Greek inscription, is parallel to the plane of the equator; the top surface is parallel to the horizon. The gnomon, now missing, is believed to have originated in the gnomon hole (where the meridian line intersects the top surface) and protruded horizontally, having its point in the axis of the cone. The vertex of the cone is above the top surface.

Photograph courtesy, Musée de Louvre, Départment des Antiquités Grecques et Romaines, Photograph by Chuzeville. For details see Sharon L. Gibbs, *Greek and Roman Sundials* (New Haven: Yale University Press, 1976), p. 268–69.

The idea of conical, spherical, planar, and cyclindrical sundials popular in Greek and Roman antiquity reappeared in sixteenth-century Europe in the form of dials using the inner surfaces of chalices, rings, and other various and sundry objects. Sundials, mostly of the disk variety, survive into our own epoch as silent and unassuming partners to their noisier and fancier kin.

The Egyptian *merkhet* consisted of a holder—a straight or forked rod—with a plumb line at its end. It has been used since early historical times for observing the transits of stars across the meridian. The instant of the transit was declared to define a certain hour of the night as the reference epoch. Nocturnal dials using similar principles told the hours of the night in Europe, twenty centuries later.

The astrolabe—from the Greek ἀστρωλάβόν*—combined in its functions everything that nocturnal dials and sundials could do. It began its history the third or second century B.C., though it did not reach its maturity until after the first millennium A.D. It retained its importance for astronomers and navigators through the middle of the eighteenth century when it was replaced by a combination of sextants and printed tables.

In its simplest form the astrolabe is no more than a device for measuring the angular separation between the directions to two distant objects. In its most sophisticated form it is an analogue computer and databank. It could be used to measure the altitude of the sun, the moon, and the stars and then, without numerical calculations, to determine the hour of the day or night, the latitude of the observer, and also to solve a number of other astronomical problems. In the Islamic world one of its many uses was to determine the times of prayer. The Arab and Latin astronomers of the Middle Ages called it the mathematical jewel, a name it well deserves. Frozen into the lines of its brass plates and into the points of its rete were such empirical data and theoretical principles concerning the motion of the stars and planets as were available to its maker. When used as a clock, it was an "interpreter" mediating between celestial and earthly processes by means of stable rules.

Earthly cycles　The oldest known nonastronomical devices used for time keeping were water clocks or clepsydrae. They employed the regularity of the rate of flow of water through an orifice, into or out of a calibrated container. Bucket-shaped, outflow-type water clocks survive from the Temple of Karnak in Egypt from around the fifteenth century B.C. Sandglasses, which are the combinations of outflow and inflow clepsydrae but use sand instead of water, may be found in fully developed forms in European paintings three millennia later, that is, in the fifteenth century.

Chinese clepsydra techniques, involving both inflow and outflow types date from the sixth century B.C. Often several water clocks

---

* Originally an adjective, *star-taking;* later a noun describing a star-taking instrument.

were used together, forming a system that was regulated by clever mechanisms of checks and balances, equipped with adjustments for the lengths of the daylight hours which changed with the seasons. Instead of water that would freeze in the winter, sand was sometimes used, as was mercury, with pipes made of mercury-resistant materials such as jade.

A peculiarly Chinese device, the incense seal clock, was already well known in the tenth century B.C. but it might have originated as long as a millennium earlier. In its most common form it is a wooden disk into which a long, continuous groove has been chiseled in the pattern of concentric rings connected as in a maze. Incense made from aromatic powders was placed into the grooves in continuous and connected segments. The incense was then lit, perhaps in the center, and began to burn slowly outward, like a fuse. The different scents identified the different parts of the day. Match cords that burned like wicks, dropping weights at intervals as the cords were consumed, have been used both in China and Japan. Oil lamp clocks were known in Europe at least from the mid-thirteenth century.

Whether it was the flow of sand or the burning of incense, a process was judged appropriate for time keeping if the clockmaker could convince himself that he was witnessing a predictable, lawful aspect of his experience of time.

Geared cycles The Antikythera machine was so named after the Greek island near which it was discovered in a sunken ship. It was a geared device, probably hand driven. It displayed the positions of the sun and the moon, it indicated sidereal and synodic months and the lunar year. It is believed to have been made during the first century B.C.

Fourteen centuries later we find geared astrolabes displaying the configurations of celestial objects by means of rotating rings and pointers. They depicted the changing positions of the planets against stereographic projections of celestial and earthly coordinates, as did all astrolabes. The relative angular velocities of the planets were mapped into the gear ratios of the toothed wheels, according to the rules of the Ptolemaic, geocentric view of the world. Thus, geared astrolabes were moving mechanical models of the universe. As simple machines, they signaled the birth of a growing family of astronomical instruments of great ingenuity. There were orreries (planetary machines) and lunaria, Jovilabes and Saturnilabes, telluria (modeling the motion of the earth and its moon around the sun) and cometaria, volvelles (that showed the tides) and armillary spheres of delicate and precise beauty.

Our mechanical clocks are also geared planetaria. The small hand follows the sun if the clock or the watch is properly positioned, albeit at twice the rate of the sun's angular velocity. The rates of rotation of the two hands are controlled by processes judged to be stable, lawful, and intelligible. The grand ancestor of their control system was the escapement of the Chinese Water-Powered Armillary Sphere and Celestial Globe Tower, described by its maker in A.D. 1090. The escapement controlled the rotation rate of a model of the heavens by means of a recurrent, cyclic program of a water wheel. Its scoops were filled with water and then regularly emptied. The Heavenly Clockwork, as it was called, was some thirty feet high. In a chamber it contained an automatically rotated celestial globe. The vast device was judged satisfactory because the motion of the objects on its celestial globe agreed with the motion of celestial objects as observed, and because of the convincing regularity of the escapement mechanism.

The earliest known European device that carried similar conviction as to its clockness was the verge-and-foliot escapement invented during the thirteenth century. The representation of the sun was simplified into the revolution of a single pointer and, later, into the revolution of two hands. From the celebrated first Strasbourg Cathedral clock of the mid-fourteenth century through Galileo's discovery of the isochronism of the pendulum in the early seventeenth century, one can trace the history of geared clockworks to the electrically controlled pendular clocks of our days. As a class of devices they sample and copy the cyclic rhythms of nature at large. They derive their authority from the intelligibility of the way in which they work and from their usefulness in our search for order, to be discussed in section 1.2.

Tabulated cycles Modern astronomical time measurement is an extraordinarily intricate process. It is not even a time measurement in the ordinary sense that someone consults a device or observes some process and then declares what time it is. It is a calculation of time whose rate of passage is determined not by celestial events in themselves but by the astronomer's understanding of those laws of nature that he believes govern the events.

First, the astronomer has recourse to the appropriate Newtonian equations of planetary motion with relativistic corrections. The solutions give the angular positions of the sun, the moon, and the planets with their moons, all with reference to the fixed stars, in terms of time. The solutions are then inverted to give time as a function of the positions of the various objects.

There is a score of time services around the globe. They are interconnected by radio links and continuously transmit time signals. These signals are controlled by local clocks, found to be reliable by earlier standards. An astronomical time observation consists of the determination of simultaneity between two instants: the crossing of the center of a telescope by a planet or star is one, the indication of the local clock is the other. Each night several such events are observed, photographed, and labeled by local clock time.

Later, using the equations of celestial mechanics, it is determined from the photographic plates how fast or slow the clock was when the object passed the reference line. The role played in this process by the equations of astronomy is identical to the role played by the designs engraved on the astrolabe: they represent our cumulative knowledge about the motion of the stars, as seen from earth. In both cases an assumption has been made: accumulated knowledge is a more reliable guide to a determination of time than the hands of the local clocks are. Accordingly, it is usually the clocks that get readjusted and not our basic theories revised.

Obviously, the determination of just when any of the events whose temporal separation is of interest did, in fact, occur, cannot coincide with the epoch of their occurrence. By the time the corrections to the clock readings have been calculated the time signals have already been beamed around the globe. Bulletins are therefore compiled, giving the revised estimates of the times when the signals were "in fact" propagated. These corrections are coordinated among many observatories and in due course second-order corrections are published. Time so determined is called ephemeris or tabulated time. From the point of view of the user of an *Ephemeris,* it is an evenly flowing time assumed to have ideally equal units, in terms of which the positions of the sun, the moon, and the planets are listed.

Let us recall the way in which the dial maker measured time. Two sets of simultaneities, remembered ex post facto, allowed him to declare that the friend-with-cheese event was separated from the bark-of-dog event by 6 or 18 arbitrary units of time, provided he divided his day into 24 equal hours, and depending on which of the two events happened earlier—as judged by his memory. Replace the two events by the crossing of the center of the telescope by two stars; replace the sundial by the local clock; conclude that when it comes to the principles of time measurement, the modern astronomer does nothing different from the dial watcher.

Atomic cycles   Atomic resonances are recurrent, cyclic phenomena similar to planetary revolutions and pendular swings. An atomic clock employs quantum transitions, that is, well-defined changes in the energy states of atoms, to generate a reference frequency. The oscillating atoms are not used themselves: they are used only to control the frequency of an oscillator and it is the output of the oscillator that constitutes the signals of the clock.

As are the cycles of planetary revolutions, the frequency at which atomic clocks oscillate is also counted, though the techniques of counting are very different.

As is the case with mechanical timekeepers, which are calibrated one against the other in terms of interpretative hypotheses, so it is with atomic clocks. They are calibrated against each other and against other cycles whose laws are believed to be known. For two of the best atomic clocks it takes about 10^{13} units of time to get out of phase by one unit. To say that this amounts to being accurate to one second in 300,000 years is theoretically true but empirically meaningless. Sampling times are hours, days, or years at the very best. If we wish to make time, as measured by atomic clocks, meaningful for long periods of time, atomic time must be referred back to ephemeris time.

The fundamental unit of time employed in astronomy used to be the mean solar second or $1/86,400$ of the mean solar day. Because of the variation in the length of the mean solar day this unit ceased to be regarded as a stable process of nature. Therefore, in 1955, the second was redefined as $1/31,566,925.9747$ of the tropical year for 1900 January 0 (which is also December 31, 1899) at 12 hours ephemeris time. This was followed in 1964 by a definition of the atomic second. It equals 9,192,631,770 periods of radiation of a particular cesium-133 transition.

Having arrived at a definition of a second from two families of different physical processes—the astronomical and the atomic—it is not surprising that the two measures of time do not generally agree and must be repeatedly synchronized according to empirical rules.

Reference epochs   The selection of a fundamental epoch is always arbitrary.

The only fundamental epoch in astronomy for which good arguments can be made is the cosmological big bang. But the numerals assigned to our own epoch with respect to that primordial instant are not particularly significant for the working astronomer. For practical purposes much more recent and much less distinguished reference epochs are used, such as the 1900 January 0, mentioned above.

The selection of a fundamental epoch for social use involves political and moral considerations and is often guided by socioeconomic conditions. Once such an epoch has been agreed upon, a calendar may be designed. Calendrical time keeping is yet another form of the friend-cheese and dog-bark story. Consider, for instance, that Sir Isaac Newton was born on December 25, 1642. This claim involves two events. Each event, in its turn, involves two instants, with each instant belonging to one of two processes. They are the calendrical process and the historical process, running parallel. The instant "birth of Newton" belongs in the historical process. It was locally simultaneous with an instant of the calendrical process, named "12-25-1642." The reference event was also the local simultaneity of two instants. One is called "the birth of Christ" and belongs in the historical process. The other is called "0-0-0" and belongs in the calendrical process.

The ideal timekeeper   Clocks and watches displaying time by means of rotating hands and a dial may be called Platonic timekeepers because they model the orderly motion of the heaven. God, wrote Plato, "resolved to have a moving image of eternity, and when he set in order the heaven, he made this image eternal but moving according to number . . . and this image we call time."[1] Digital displays may be called Aristotelian. We measure more or less by number, reasoned Aristotle, and we measure more or less movement by time. More or less movement is more or less time; therefore "time is just this, the number of motion in respect to 'before' and 'after.' "[2] Digital clocks count the number of motion, the human observer provides the specifications for "before" and "after."

Superimposed upon such historical continuities as just implied, we find changing ideas as to what constitutes an ideal clock. In his *Horologium Oscillatorium* (1673) Christiaan Huygens wrote that with the help of geometry he had discovered a wonderful new way of measuring time.[3] He believed that his cycloidal pendulum could be used to control an ideal clock because he was able to demonstrate, theoretically, that the pendulum remained isochronous regardless of the amplitude of its swing. A simple Galilean pendulum was known to be isochronous only for small displacements.

The ideal clock appropriate to the physics of our age is called the Einstein-Langevin clock.[4] It is a purely theoretical device consisting of a pair of parallel mirrors attached to a rigid body. A light ray is imagined traveling back and forth between the two mirrors. The instants to be counted are the arrivals of the light ray at one or the other mirror. This device fits the concept of an ideal clock because the clock is controlled by the principles that also govern the whole of kinematics and dynamics, namely, relativity theory. Consequently, its functions are in complete harmony with the rest of our understanding of the macroscopic universe just as Huygens's cycloidal pendulum functioned in complete harmony with his interpretation of the world. Should our comprehension of the universe change, so will the form of the ideal clock.

The notion that a continually improving accuracy of clocks makes sense implies a particular view of time, hidden by convention. What is being claimed is that the improved device indicates more truly the rate at which time passes than the prior devices did. Evidently, therefore, the rate at which time passes cannot itself be controlled but only recognized and recorded with increasingly greater precision, using improved methods of time measurement.

In 1922 Einstein remarked that "we are still far from possessing such certain knowledge of theoretical principles as to be able to give exact theoretical construction of solid bodies and clocks."[5] A sufficiently broad view of timekeepers, as sketched in this section, suggests that any process may be used for keeping time as long as its regularity is convincing, judged in the context of our total understanding of nature. Therefore, it will never be possible to formulate principles leading to the construction of a final, ideal clock. But it is possible to develop principles leading to improved understanding of physical—and biological and behavioral—processes and then select any one of the innumerable, possible processes as the one favored for time keeping. The reasons for the selection may be pragmatic, aesthetic, or even ethical.

As to the construction of a solid body or rigid rod, Herman Bondi has argued:

Not only do we know that, by using radar methods, all lengths can be measured in time, but in fact the size of our measuring rods is determined by atomic interactions which themselves are fully characterized in principle by atomic frequencies. Thus, only time standards are true primary standards, and units of length are determined from them by the use of c. . . .[6]

Salecker and Wigner argued, on a different basis, that both spatial and temporal intervals ought to be measured by clocks, using the appropriate instructions of relativity theory. They maintained that rods are necessarily macroscopic objects that will interact with their environments in uncontrollable ways, whereas clocks may remain microscopic objects.[7] On either Bondi's or Salecker and Wigner's count, time measurement has a priority to distance measurement. This thought leads us back to questions of time keeping in general.

1.2 The Search for Order

A careful, comparative review of the methods sketched in the preceding section suggests three conditions as having been met for each of the examples given. I will assume that these are necessary as well as sufficient conditions for time keeping and will refer to them as the rules of time measurement.

(1) It is necessary to recognize two continuous processes, each with identifiable, distinct states; that is, each capable of defining countable instants.

(2) It is necessary to recognize two nonsimultaneous events, with each event comprising a local coincidence of two instants, one instant each from the two processes.

(3) It is necessary to have a rule of transformation, or at least to assume that such a rule exists, whereby the number of instants between the two events counted along one process may be changed into the number of instants counted along the other process.

The first and second rules, taken together, tell us that no process in itself is sufficient to measure time. If I say that I stop for lunch at 1 P.M. I describe the local simultaneity of two instants. One belongs in a process known as the United States Time Service, and is recorded by the states of my electric clock. The other instant belongs in a process known as my life, and is indicated by the state of my hunger. By analogy, I may look at my straight edge and say that it shows 7 inches, which it does. It also shows 2½ inches and

12 inches. I shall have made a distance determination only if I said that my jade plant was 7 inches tall. In the case of the length measurement, the need to have two distances before a comparison can be made is banally obvious. In the case of time measurement, the need for two processes tends to be hidden by convention.

The third condition is of very great significance because the rules of transformation it demands are the laws of nature.

Let us turn to the practical question of accuracy.

Imagine that we are monitoring the running patterns of rats. The plot will show circadian rhythms of impressive regularity and reproducibility. But we will still trust the laboratory clock more than we trust the rat clock. This is the reason why the circadian rhythms of rats are always plotted against laboratory clock indications as the independent variable, and not the other way around. The rat ticks are unpredictable to a much greater degree than the clock ticks are because we understand the functions of the laboratory clock much better than we understand the operation of the rat clock. Were the case the inverse, we would calibrate our laboratory clocks against the running pattern of rats. Both the rat clock and the laboratory clock are synchronized to the sun but the rat is a more complicated system. Therefore, it is less predictable; that is, less accurate.

We learned that the ephemeris and atomic seconds are not identical in length, warranting occasional adjustments. The adjustments always consist of changing the number of seconds in a year. We say that the year got longer and not that the atomic second got shorter, because the atomic clock is judged to be a more reliable device than the earth clock. If it were the other way, physicists would immediately set out to discover why atomic clocks vary their rates, using the uniform rotation of the earth as a reference.

As said above, the best atomic clocks are stable to one part in 10^{13}. On a daily basis this is an instability of 10^{-8} seconds. What would we have to know about the earth before its rotation rate could be calculated from first principles to comparable certainty? We would need precise theories of tidal motions, crustal changes, and winds, of weather patterns, the behavior of the mantle, the outer and inner cores, and surely of other details of which we may be quite unaware. With all the theories worked out it would still be necessary to know the boundary conditions of each variable to appropriate accuracy.

We have no ways for calculating earth-clock indications to accuracies comparable with those of the atomic clock, or for calculating rat-clock indications with accuracies comparable to those of the earth clock, obtained empirically. Neither could we measure the boundary conditions.

The atomic, astronomical, and biological clocks form three distinct families of timekeepers. Clocks belonging in the same family remain comparable in degrees of unpredictability or, in practical terms, in degrees of accuracy.

For instance, physiological clocks are sufficiently precise to make the ecology of living systems possible. But we have no first principles whatsoever from which their rhythms could be calculated. Astronomical cycles are more precisely predictable from a combination of empirical data and the equations of celestial mechanics. Atomic clock readings are predictable with phenomenal accuracy, using equations derived from first principles and a few constants of nature, determined empirically.

For the different degrees of predictability from first principles, I wish to assign the responsibility to a variable called *complexity*. Giving a definition of complexity is very difficult in spite of the fact that the notion seems intuitively obvious. We will work our way toward a definition, nevertheless, and even toward a mode of quantifying complexity.

Imagine now that we are measuring a particular physiological rhythm of a laboratory animal. Based on theoretical knowledge and on experimental conditions, we make a prediction about its phase at 12 o'clock noon, in terms of the laboratory clock, a week from today. But when the time comes, the rhythm will be found out of phase with its predicted value. The task of the biologist studying the rhythms will be to account for the difference between the predicted and observed values. If all experimental variables of which she knew were carefully controlled, she will look for biological processes thus far unknown and will hope to discover new laws of nature. Eventually, she plans to have all the equations necessary to change biological clock readings (of the selected processes) into laboratory clock readings.

Replace the pair, biological clock–laboratory clock, with a pair of identical atomic clocks running side by side. Ideally, the transformation equation between their readings should be an identity. But after having run side by side for 300,000 years, they are found to be off by one second. The new atomic clock that was being tested should have ticked 10^{13} times. Instead it ticked 10^{13} +1 times. The task of the physicist will be to account for the difference between the predicted and the observed values. If all experimental variables of which he knew were carefully controlled, he will look for some new laws of nature. Eventually, he plans to have an equation that can be used to change either clock reading into the other one.

Next, instead of letting the two identical clocks run side by side, let one of them be taken on a journey. Imagine that upon its return the traveling clock will have recorded 6,789 repetitions of a certain physical state (6,789 instants) whereas the stay-home clock marked 9,876 instants between the same two events. Which clock is correct?

The situation is not different from the case when the clocks were out of phase by 1 part in 10^{13}. This time they are out of phase by 3×10^3 parts in 10^4. If the difference can be accounted for by a suitable transformation equation then a time measurement will have been performed and both clocks may be judged correct. If no such rules are known we ought to look for them and, when we find them, we shall have a new law of nature.

The statement emerging from these deliberations is a new answer to an old question: What do we do when we measure time? We test the validity of transformation equations. More generally, we search for order among clock readings and, when found, we express them in laws of nature.

What any particular clock or group of clocks indicates about time says no more about time than the lengths of different rulers say about the nature of space. Scientific knowledge regarding the nature of time may be found, instead, in those rules that transform clock readings.

In a fine essay, G. J. Whitrow has examined the nature of the scientific method, "a curiously elusive problem." He summed up his findings as follows.

> The question of existence or non-existence of "scientific method" can be answered affirmatively, not by recommending some philosopher's stone that can turn mental base metal into gold, for example the "inductive method" or the "hypothetico-deductive method," but by careful analysis of the major steps by which the sciences, and in particular physics as the most highly developed science, have in fact advanced. In this way

. . . we see that the history of the scientific method can be regarded as the evolution of a hierarchical succession of "orders of questioning." Three general factors are involved: the production of *empirical evidence,* both observational and experimental; the exercise of *imagination* in formulating specific questions and hypotheses, and scientific *judgment* or insight. . . .[8]

The findings of this chapter suggest that scientific knowledge about time cannot be found in clock readings (the "empirical evidence") but is implicit, instead, in the laws of nature (the "exercise of imagination"). How may the knowledge implicit in physical hypotheses be made explicit? By offering "scientific judgment or insight" in terms of a coordinated interpretation of the time-related teachings of modern science.

The Principle of
Temporal Levels

The conclusions of section 1.2 opened the way toward the formulation of a new theory of time. It did so by recommending a coordinated interpretation of the time-related teachings of contemporary science. As a guide to the task we will follow the instructions of Newton given in Book 3 of this *Principia:*

> In experimental philosophy we are to look upon propositions inferred by general induction from phenomena as accurately or very nearly true, notwithstanding any contrary hypotheses that may be imagined, till such time as other phenomena occur, by which they may either be made more accurate or liable to exceptions.[1]

The natural phenomena of our interest are the temporalities identified by the major theorems of modern physics. The "propositions inferred by general induction from phenomena" are the tenets of the principle of temporal levels, sketched in this chapter and developed in this book. Those propositions amount to a new view of reality. Section 2.4 examines certain "contrary hypotheses." Section 2.5 tells us what may be meant by "other phenomena" in our particular inquiry.

2.1 The Moving Boundaries of Reality

Early in this century, the German biologist Jakob von Uexküll drew attention to the fact that an animal's receptors and effectors determine its world of possible stimuli and actions and hence the extent and nature of its universe. He called such a species-specific universe the *Umwelt* of the species.[2]

Von Uexküll was concerned with the functional cycle of the

animal. This cycle is an information carrying loop. Along the loop external signals pass and become internal signals. They are processed and evaluated in terms of the capacities of the organism and the modalities of its functions. Then they are passed on as commands that control, more or less, the animal's behavior. Through the consequent behavior the external signals themselves become modified. The umwelt of the species can have only such characteristics as may be processed by the functional loop. In terms of modern semiotic theory, organisms do not perceive a final and absolute reality but only signs that have been filtered and formed by their specific capacities.[3] From these signs they construct what, for each species, is the reality of the world.

If animals could write, their physics would be void of many concepts we regard as essential for a valid description of the universe. Also, their sciences would contain claims about structures and processes that we would have great difficulty in recognizing. Both space and time would appear in strange guises.

A corollary of what I have just described, following von Uexküll, is that the umwelt of a species, for each member of the species, is a self-consistent and complete world. It is all there is or can be to the universe. Whatever is external to the umwelt must be taken as nonexistent from the point of view of the organism. For instance, certain butterflies have patterns on their wings which show up in ultraviolet light. These patterns exist for other butterflies but not for vertebrates because vertebrate eyes are not sensitive in the ultraviolet region.

Consider next the spider's web. It does more than catch flies. It is a sort of polar coordinate, an extension of the spider's sense: from its vibrations the spider can tell the size and location of its prey. Flies not caught do not exist for the spider. This eight-legged relative of the insect would have little use for a Cartesian space of three dimensions; its space has only angles and distances. Think next of the fact that all nocturnal animals and all cave dwellers live primarily in tactile space. Insofar as tactile sensations do not diminish with distance as visual sensations do, the objects of tactile space do not grow smaller as they get further away. Instead, they suddenly vanish. The geometry of tactile space must be different from the geometry of the space of our experience, which is tactile and visual in its near field and visual in its far field.

Likewise, time as tempo and time as duration will vary from

species to species depending on the species' complement of receptors, effectors, and rates of metabolism.

"Ask any molecule," wrote John A. Wheeler recently, "what it thinks of the second law of thermodynamics and it will laugh at the question."[4] The figure of speech is well taken. If we think of the umwelt of the individual atom as comprised of those forces to which it is able to respond, in that umwelt the second law of thermodynamics can be given no meaning. As far as the individual atoms are concerned, that famous law does not exist. It is a statistical statement meaningful only for the umwelt of collectives of atoms.

In modern psychology *umwelt* is defined as "the circumscribed portion of the environment which is meaningful and effective for a given animal's species. . . ."[5] Members of our own species are no exceptions to von Uexküll's umwelt theory. We also have our functional loops. The "environment" of which the animal's umwelt is a "circumscribed portion" is our own human umwelt. Its biological boundaries are determined by the sensors, processors, and effectors of our functional loop. But the human umwelt has some peculiar features. Our spider's web includes an impressive array of exosomatic devices, such as the tools of industry and the instruments of science. Although we are vertebrates, we may, nevertheless, include in our umwelt the ultraviolet patterns of butterfly wings because we can photograph them, then display them in that portion of the electromagnetic spectrum which we experience as visible. Through creating and manipulating the symbols of mathematical physics, we may even learn about the history and structure of our universe and about atomic structures and processes, not directly open to the senses.

It has been a cardinal assumption of physical science that the properties of natural processes and things are independent of us and, therefore, physics can deal with truly objective aspects of the natural world. The tremendous success of physics and of its offspring, industrial technology, has been taken to demonstrate that the teachings of physics transcend historical and philosophical biases and reveal, therefore, a value-free and man-independent reality. These beliefs are very questionable but they make for a satisfactory idealization. At least they do so until one begins to seek the outlines of certain universals, such as an understanding of time, shared by all branches of physics, by other scientific disciplines, and by the humanities as well.

G. J. Whitrow spoke for the generally held view in natural
philosophy when he asserted that time is a single unitary feature
of the universe, "although certain aspects of it become increasingly
significant the more complex the nature of the particular object or
system studied."[6] If this is the case, it should be possible to trace a
continuous path from the time—or absence of time—in the world
of the atom, to the time of inorganic matter, living organisms, and
society. But contrary to popular misconceptions, no such path has
ever been traced—nor did anyone ever tell us why it is difficult
or perhaps impossible to trace one, within the natural order of
things.

As a step toward an integrated understanding of time in which
that path can be traced, a new theory of time will be sketched in
section 2.3.

The arguments leading up to it begin with the umwelt principle. Following von Uexküll's assertion that the umwelt of a species
is its universe of reality, the human umwelt is identified with the
completeness of human reality. The world revealed through experience, experiment, and theory is the way the world is. But the boundaries of this totality are movable. There are new realities to be discovered and this surely will always be the case. But the features of
a yet unknown world could hardly be classed under the heading
reality. They must first become knowable and known.

Using the wisdom of hindsight, it is not surprising that von
Uexküll's rich notion first arose in the context of biology. It began
with animals and man because methods for the exploration of the
umwelts of infants, apes, and the tick (von Uexküll's exemplary
species) were easily available. We knew how to put questions to
nature via the dog, the frog, and the earthworm and thus construct
the outlines of their umwelts. But physicists have been doing the
same for elementary particles and galaxies. They can determine the
rules that do, and do not, apply to those universes. We can put
questions to nature via the accelerating electron and the receding
galaxy. Let us imagine, therefore, that it makes sense to think about
the umwelts of nonliving objects.

It is obvious that a world of distinct and only partially over-
lapping umwelts—of young children, birds, viruses, molecules—are
still features of *our* umwelt. By the reasoning already given, they
may be taken as real only to the extent that we can know and
comprehend them. There is nothing inconsistent about the situation.
It is the conscious human mind that searches for order among in-

organic and organic phenomena, then writes natural philosophies about them; it is not the other way around.

It is we who are capable of thinking ourselves into the position of radiation, particles, or field mice, and of outlining the boundaries of their universes to the limits of our capacities. With these limitations in mind, we must regard those universes as complete in themselves. This extension of the originally biological notion of the umwelt is the generalized umwelt principle. We will find that it introduces a great deal of economy into our thoughts about time in scientific theory and experiment. For it will enable us to determine whatever information is irrelevant, and allow us to discard it.

The significance of the generalized umwelt principle, however, goes much beyond issues of methodology. It points in the direction of a new perception of reality. The new understanding is warranted by the fact that we can recognize in nature a hierarchy of distinct umwelts, associated with a hierarchical ordering of the structures and processes of the inorganic and organic worlds.

2.2 The Stable Integrative Levels of Nature

The modern concept of integrative levels, used in organizational theory, is a latter-day version of the much older concept of hierarchical ordering.

That the universe is arranged in different orders of being, descending from and ascending to a first principle, is an idea going back at least to Plato and Aristotle. The term *hierarchy* first appears in the treatises of Pseudo-Dionysius, a Christian Platonist of the second half of the fifth century. In *The Celestial Hierarchy* and *The Ecclesiastical Hierarchy* he maintains that it is the hierarchical organization of the universe that makes it possible to reconcile in its nature the phenomena of the many with the unity of the one. In the medieval view of the world, the universe was perceived as a graded chain of beings, stretching forth from the Deity down to the least perfect being on earth, with the organization of the Church of Rome an image of the ordering.

Protestant reformers and early modern scientists rejected the hierarchical concept of the universe and tended to support the idea of a cosmic rule in which all beings partake equally. The Aristotelian heavenly spheres and their later angelic complements were abandoned for cosmic democracy, giving Voltaire a chance to write that there could be no continuous, hierarchical chain of living beings in the universe because there was no gradation among the heavenly bodies.

In our own day some maintain that the history of the concept of hierarchy is the history of Western obsession with order.[7] Others reject it as a projection of social stratification promoted by the ruling classes of the capitalist world. Official Marxist philosophy maintains, however, that hierarchically organized forms exist in all spheres of objective reality: the inorganic, the organic, and the social, and that within Marxist philosophy, "the idea of hierarchy of qualitatively irreducible structural levels has been developed."[8]

The concept of hierarchy enters mathematics through Russell's ramified theory of types. Here the hierarchy is built up from levels wherein members are defined in terms of the totalities of objects of the lower level. More recently it appeared in catastrophe theory where objects of a codimension have certain properties in common, including that their descriptions require the same number of equations, as distinct from objects of higher or lower codimensions. All these formal systems deal with hierarchies of increasing complexities.

The hierarchical structuring of biological systems is a familiar fact. It has been worked out in detail by J. H. Woodger[9] who axiomatized it and introduced into it a degree of rigor, by defining its concepts within the study of organismic biology. He noted that a member of any level may always be regarded from three points of view: from that of its own membership, from that of a member of a higher level, and from that of the next lower level. He also stressed the necessarily different kinds of lawfulness of the different levels and maintained that the laws determining the behavior of the parts of a given level of organization leave undetermined some aspects of the behavior of the parts of the higher level.

Herbert Simon has given reasons why hierarchical structuring is the most efficient way of handling complex systems and noted that in the course of evolution complex systems do, in fact, tend to evolve hierarchical ordering.[10] If anything goes wrong, the system does not have to be started again from its simplest elements. The grouping of structures along hierarchically nested levels of increasing complexity produces stability points. They make it possible for the next level of organization to emerge, because the lower levels do not get easily undone. In all cases, however, the issue is not one of simple aggregation but aggregation with communication.[11]

In general organizational theory, hierarchical control is seen as arising from internal constraints that force the elements into collective behavior, independent of any detail of the behavior of the elements themselves. An example, the second law of thermodynamics, has already been given. Cyril Stanley Smith, writing about the structural hierarchy of inorganic systems in general, remarked that connected disorders on one level of organization tend to give rise to significant patterned behavior on a higher level.[12]

Joseph Needham in his Herbert Spencer Lecture[13] addressed the theme of hierarchical integrative levels in nature: of successive levels of organization, of successive forms of order along a scale of complexity. He noted that the assumption of integrative levels necessarily demands, as its corollary, the existence of mesoforms, of structures and functions that would fall between two integrative levels. H. H. Pattee, making the same observation, remarked that each side of an interface between integrative levels "requires a special language. The lower-level language is necessary to give what we might call legal details, but the upper level is needed to classify what is significant."[14]

If it is indeed the case that the universe is organized along distinct and stable integrative levels then such a division ought to be identifiable in the ways knowledge has become organized. For instance, the major theorems of physics should show preferences for dealing with the structures and functions of an integrative level and an inability to deal with—or an irrelevance to—the structures and functions of another integrative level. The fields of scientific knowledge in general and the great theorems of modern physics in particular, could then be arranged in an order corresponding to the integrative levels. One may even expect that the further apart along this scale any two principles or sciences are located, the more stubbornly they would resist being combined on equal footing under a single heading.

The compartments of contemporary science and the major theorems of modern physics do display exactly such properties. Consider the array of Table 1, page 26.

The first and most obvious distinction that may be made among the three physical theories relate to the dimensions and the variety of objects with which they deal.

We will see in chapter 3 that special relativity theory succeeded in tying kinematics and electrodynamics to the motion of light. The far-reaching consequences of the theory follow from its fundamental concern with the nature of electromagnetic radiation.

The central object of its interest is the photon: a curious creature possessing a unique quality of motion and an immense speed. As a particle it is dimensionless and has zero mass. The class of such objects is limited to three: photons, gravitons, and neutrinos though maybe only to two—photons and gravitons.

Quantum theory is the fundamental theory of matter. Its concerns and formalism include the physics of light but its natural focus is the interaction of matter with radiation. Its central objects are atomic and subatomic particles. Quantum theoretical considerations are significant for all members of the particle zoo including the two or three massless objects which constitute its lower boundaries. As aggregates of particles reach certain dimensions, large by atomic standards but quite below the size of the smallest living organisms, quantum mechanical effects become generally either undetectable or unimportant.

General relativity theory is a theory of gravitation. Its natural focus is the collective behavior of particles gathered into the vast ponderable masses of stars, the stars into galaxies, the galaxies into the object known as the physical universe. It governs the behavior of atomic aggregates but not that of atoms and subatomic particles. Although in principle the domain of general relativity theory includes everything from neutrinos to apples to galactic clusters, its most peculiar results deal with the world of the cosmologist.

How do combinations of these three theorems fare in regard to the ease of their unification? We will now examine such combinations with reference to the relative distances between the theorems shown in Table 1.

The combination of special relativity theory and quantum theory has yielded rich results, including the discoveries of the atomic spin and the existence of antimatter. Relativistic field theory, a com-

Table 1. The major theorems of physics are arranged according to the integrative levels of nature with which they are primarily concerned. The two headings on the right are shorthand notations entered here to help maintain perspectives upon physical theory.

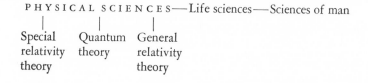

PHYSICAL SCIENCES—Life sciences—Sciences of man

Special relativity theory   Quantum theory   General relativity theory

bination of the two theorems, has been successful in describing the interaction of electrically charged matter with electromagnetic radiation.

Turning from next-door neighbors to neighbors-once-removed, special relativity theory is, formally, a limiting case of the general theory of relativity because it describes the world without matter and gravitation. All statements derived from the general theory must be locally special relativistic. Although there are general relativistic effects in the local domain of astronomical phenomena and there are special relativistic effects in the same domain, the two theorems do not have an obvious common focus. Thus, special relativity theory is readily and necessarily invoked for problems in atomic physics while general relativistic effects are negligible or totally absent. For its part, general relativity theory is necessary in dealing with the large-scale behavior of a nonempty universe while special relativistic effects remain of local, that is, galactic interest. Formally, the main features of the most important general relativistic models of the universe are identical with those one could predict from the application of Newtonian physics to a non-Euclidean model of the world, whereas special relativistic teachings cannot be derived from Newtonian physics, even in a poor approximation. Fock pointed out long ago that the name, general relativity theory, is a complete misnomer because "what is actually generalized by this theory is a hypothesis concerning the nature of space-time, and not the concept of relativity."[15]

The combination of general relativity and quantum theory (see Table 1) has been frustrated by profound technical and conceptual difficulties. In the words of Hawking and Israel, "at the moment it is not clear what form a consistent quantum theory of gravity would take or whether one exists at all."[16] The separation between the two theories is, of course, not absolute but only very pronounced. Thus it has been shown that the presence of gravity can give rise to quantum effects. But the kind of relations that have thus far been identified are significant only for such extreme conditions as are believed to prevail during cosmological particle production and black-hole evaporation, near the chaotic boundaries of the structured universe.

Consider now not only physical science, but modern science in general, and once again with reference to Table 1. It is difficult to think of ways in which peculiarly special relativistic effects enter the life sciences. Quantum theory does become important in the new and difficult field of molecular biophysics via some of the methods used in understanding the molecular structure of living matter: infrared spectroscopy, ultraviolet spectroscopy, X-ray diffraction, and electron microscopy. But biologically important molecules are immense and hence the work of the molecular biophysicist tends to be closer to that of the chemist than to that of the quantum mechanicist.

Do general relativity theory and biology, next-door neighbors in Table 1, have any common interest? At first sight one might not think so, but reflection reveals that they combine with ease. The locally satisfactory form of general relativity theory, a theory of gravitation, is Newtonian dynamics and kinematics. All biology is played out in the gravitational field of the earth. An analogous link between special relativity theory and biology cannot be found because the peculiarities of the special theory become evident only at speeds near to that of light, and such conditions are alien to living matter.

Twice removed from physics in Table 1 are the sciences of man: psychology, anthropology, and sociology. Though no living organism or human civilization can do without matter, the concerns of the physical sciences are far removed from the sciences of man. Their obvious kin is biology.

The grand unification of the major theorems of physics, the ones listed as the first three entries on the left of Table 1, has thus far eluded the best of minds. The theorems have been selectively combined in ways already mentioned and other ways are continuously being sought. But no universal theory is within sight to combine with ease and self-evidence—hallmarks of a good theory—the physics of light, of particles, and of the universe, on equal terms. The difficulty may reside in the relative independence of the different organizational levels of nature that, by necessity, is reflected in the independence of the theorems. We will return to this issue in section 6.5.

Presently it is proposed that six major integrative levels be distinguished.

(1) The world of particles with zero restmass, always on the move at the speed of light.

(2) The world of particles with nonzero restmass, always on the move but at speeds below that of light.

(3) The world of massive, ponderable masses gathered into stars, galaxies, and groups of galaxies.

(4) The world of living organisms.

(5) Man as a species and as an individual member of the species.

(6) The collective institutions of human societies to the extent that they function as semiautonomous structures.

It is postulated that these levels of organization are distinct, stable, and bear a hierarchical, nested relationship to each other. *Distinct* means that structures and processes belonging in an organizational level differ from those belonging to another level more significantly than members of an organizational level differ among themselves. *Stable* means that the integrative levels are composed of structures which nature prefers to maintain for extended periods of time. Mesoforms are rare and unstable (sec. 7.6). *Hierarchically nested relationship* among the stable integrative levels means that the structures and processes of each level subsume those of the level(s) beneath them, are restrained by the lower-level regularities, while also possessing certain new degrees of freedom of their own.

2.3 The Canonical Forms of Time

Let the conclusions of sections 2.1 and 2.2 be combined in a single postulate: the reality in nature of a hierarchy of nested, level-specific umwelts. Next, by the freedom of "scientific judgment or insight" offered and demanded by the scientific method, let the reality of level-specific temporalities be inferred.

What the hallmarks of these level-specific temporalities may be will be explained by means of a visual metaphor: an arrow drawn on a sheet of paper.

We begin by imagining a well-defined arrow: head, shaft, and feather. The picture stands for the temporal umwelt unique to the human mind. It is a time informed of sharp division between future and past, of long-term expectation and memory, and of a *mental present* with continuously changing boundaries. These are the hallmarks of the *nootemporal* umwelt. The term was coined from the Greek νόος, which means mind or thought. The noetic umwelt is created by our capacity to produce symbolic transformations of experience and then manipulate them as part and parcel of reality. The features of this umwelt are determined by its processes and structures: the arts and artifacts of civilizations including language, industry, and the theoretical and experimental means of science.

Let the head and the tail of the arrow become ill-defined,

amounting to no more than ambiguous but still distinct limits to the shaft. The picture is a visual metaphor for *biotemporality*. This is the umwelt of all living organisms (including man, as far as his biological functions are concerned). It is also the umwelt of certain fantasies and dreams.

Living species display a very broad spectrum in the degree to which they define their distinctions between future, past, and present. But biotemporal arrows remain quite different from noetic ones. In the biotemporal world the mental present of the noetic umwelt reduces to the *living* or *physiological present*. In its highest reaches this is a category of the perceptual and cognitive set but without mental content.

The mental present of the noetic and the physiological present of the biotemporal umwelt are two forms of nowness. It is with respect to nowness, or presentness, that future and past acquire meaning. By the hierarchical, nested organization of nature the mental present of man subsumes his physiological present and must be assumed to have evolved from it.[17]

The functional basis of the physiological or living present is the necessary internal temporal coordination of living organisms. The physiological present is the phenomenological witness to the *simultaneities of need* which must be maintained if the autonomy of a living organism is to be assured. In the case of man, the physiological present has given rise to, and remains coexistent with, the mental present, characterized by *simultaneities of intent*. More will be said about these phenomena.

The head and tail of the metaphorical arrow may even be totally absent. What remains is the shaft of the arrow, a line, an image of *eotemporality,* so named after Eos, the goddess of dawn. This is the time represented by the physicist's t in equations usually described as not responding to the direction of time. This is the time of the macroscopic aggregates of matter, making up the astronomical universe.

The eotemporal world does not contain structures capable of protecting and maintaining their identities as living organisms do. Nothing among eotemporal functions can correspond, therefore, to simultaneities of need or intent. It follows that in the eotemporal world no meaning can be attached to the concept of now. Coincidences do occur but they must remain *simultaneities of chance.* As futurity and pastness make sense only in terms of a present to which they are referred, it further follows that eotemporality is a directionless time, one of *pure succession.* It is a completely symmetrical time. Purely cyclic processes are eotemporal as is an ideal clock. It is eotemporality that Eddington implied when he remarked, "the more perfect the instrument as a measurer of time, the more completely it conceals time's arrow."[18]

The shaft of the arrow may disintegrate into slivers of wood, into geometrical points representing disconnected fragments of time. The image stands for a world where it is impossible to give concrete instants either theoretical or empirical meaning. Temporal positions may only be specified probabilistically. The reader will recognize the universe of atomic and subatomic particles. Such umwelts are named *prototemporal* for proto-, the first of a series.

Finally, even the remaining fragments of the metaphorical arrow may vanish. A blank sheet of paper is left, a symbol for an *atemporal* umwelt. Atemporality is not to be mistaken for the philosophical idea of nothingness. It might better be associated with the pre-Socratic notion of Chaos, a state of affairs which was said to have preceded the emergence of the world. Imagine a universe composed entirely of particles that travel at the speed of light. It is known from special relativity theory that in the proper framework of the photon no meaning can be attached to ideas of futurity, pastness, and presentness (as the state separating future from past). Such a universe is atemporal. Some of the early Greek cosmogonists identified Chaos with Tartaros, a sunless abyss and the lowest part of the underworld. We identify it with the most primitive and the primordial integrative levels of the universe.

In dynamics, canonical equations mean the most significant and simplest forms to which other equations may be reduced, without loss of generality. In the same sense, the temporalities just described may be called the canonical forms of time. They are first encountered as symptoms of complexity in the dynamics of the different integrative levels. But they also appear in the dynamics of all higher integrative levels. This is the kind of relationship that is meant by the nested, hierarchical organization of nature.

2.4 "notwithstanding any contrary hypotheses that may be imagined"

In 1908 the Cambridge philosopher J. M. E. McTaggart published a paper in which he distinguished what he called an A-series and a B-series in time. By *A-series* he meant that idea of time which involves futurity, pastness, and presentness. This, he maintained, "is a constant illusion of the mind [as] the real nature of time contains only the distinction of the *B-series*—the distinction between earlier and later."[19]

Though not always under the names of A-series and B-series, McTaggart's distinction found ready acceptance in the literature of time in physics. It represents a view of time that physics appears to teach. But, as may be easily shown, the concept of the B-series—a condition of earlier-later but without a now—is self-contradictory. For that reason, it must be rejected.

Figure 2 shows two nonsimultaneous events: *FP* (for Frog-Princess) and *FF* (for Frog-Frog). Consider *FP* as being earlier than *FF* (or *FF* as being later than *FP,* which is the same). This is an example of McTaggart's conditions for the B-series. But it is a sleight of hand.

If I say that *FP* is earlier than *FF* I have hidden the "now" at *FF* and placed *FP* into its past. If I say that *FF* is later than *FP* then I have hidden the "now" at *FP* and placed *FF* into its future. (The same kind of reasoning holds had I said that *FP* is later than *FF* or *FF* earlier than *FP*). But the B-series, by definition, can contain no present at *FF* or at *FP* or anywhere else. It follows that the world of the B-series cannot be nowless and yet permit conditions to which the relationship before-after can apply.

If time is without a "now" and is also assumed to be continuous, it can only be the pure succession of the eotemporal world.

The variable *t* in those equations of physics that are said not to respond to the direction of time represents eotemporality. Natural languages have no words to describe dynamic changes in a world of pure succession. This is the reason why a new term had to be coined.

It is also necessary to comment on McTaggart's phrase, "a constant illusion of the mind." Presentness and, with it, futurity and pastness are not introduced into the nowless substratum of the physical world by the functions of the human mind. Rather, they come about through the dynamics of the life process, without any reference to the mind. Furthermore, by the generalized umwelt principle, future and past are not illusions of the mind or phantom sensations of the body. They are part and parcel of the temporal reality of processes and structures above the physical ones.

FP

FF

Figure 2. Issues of before-and-after. An illustration to McTaggart's B-series. Courtesy, Recycled Paper Products, Inc. All rights reserved. Design by Sandra Boynton. Reprinted by permission.

The concept of time asymmetry has been universally employed in the literature of time in physics. P. C. W. Davies explains the idea clearly and concisely.

> The two directions of time in the following sense—*towards* the past and *towards* the future—are known from experience to be fundamentally distinguishable physically. This fact is quite independent of the existence or motion of the now. For example, we *remember* the past. Moreover, this *asymmetry* with respect to the two time orientations is also readily recognized in the laboratory.[20]

Time asymmetry so understood is another name for McTaggart's B-series. In fact, it invokes the B-series twice: once going into the past and once going into the future. For reasons already given, this kind of time concept is self-contradictory. The direction of time recognized in the laboratory, in the cathedral, in the museum, in

Calypso's bedroom, everywhere where there are living organisms, is always referenced to a present. One cannot have asymmetry without a now. What one can and does have, instead, is the time of pure succession or eotemporality. But eotemporality is not asymmetrical at all.

A distinction between the time of physics and the time of everything else is often made by an epistemological dichotomy. Again, Davies puts it very well.

> Although we are forced to conclude that the laws of physics do not themselves provide asymmetry, it is one of the most fundamental aspects of experience that as a *matter of fact,* the world is asymmetric in time. This is sometimes expressed by saying that the temporal asymmetry is "fact-like" rather than "law-like" or "extrinsic" rather than "intrinsic."[21]

The distinction very clearly implies that it is only through physical science that the lawlike aspects of temporal passage may be established. Should the directional nature of time be found to have its roots in the life process and in mental functions, then both the biological and the behavioral sciences would become factlike but not lawlike propositions as far as time goes. The contrary situation must be asserted: the different temporalities have lawlike aspects precisely because, for carefully defined conditions, they are factlike. Scientists generally pride themselves in building their laws on facts. Note further that extrinsic and intrinsic must have a reference. That reference is either nature, or physics, or the two somehow equated with each other. One must wonder where and how life, mind, and society fit into the scheme.

The literature of time in physics abounds in hypotheses believed to reconcile the temporal behavior of physical systems, expressed in the equations of physics, with an unanalyzed concept of "laboratory time." Admittedly, the issues are subtle. The three examples given represent the most common absences of careful criticism.

2.5 "by which they may either be made more accurate or liable to exceptions"

Interpretative propositions when they aspire to cover large areas of knowledge are usually called principles. The idea of level-specific temporalities has thus far been an inference, a conjecture. Let that inference be raised to the status of a postulate and let it be called the *principle of temporal levels*.

It is a generally accepted hypothesis of physical cosmology that the universe began from a primordial state of chaos. In subsequent stages emerged the particulate form of matter, followed, as the universe cooled, by the formation of stars and galaxies. On a speck of matter known to us as the earth, life was born, then man and man's social institutions. The stable integrative levels created by these steps survive and coexist today. It follows that time itself has evolved along a path corresponding to the evolutionary complexification of matter.

This notion must appear as strange as the idea of organic evolution by natural selection must have appeared to most of Darwin's contemporaries.

There are good reasons to believe that the acceptance of any of the temporalities below the biotemporal as an aspect of the dynamics of a world, complete in itself, encounters the difficulty of regressive sharing. Imagining a world in which time is without an arrow, or even without asymmetry, is inhibited by those functions of the mind that protect the integrity of the self. We have a built-in resistance against monkeying with time, a kind of biological chauvinism. But this is a useful prejudice. It is rooted in the need to secure our integrity as humans vis-à-vis the nonhuman forms of life. In any case, consistent with this policy, natural languages tend to prohibit assertions about the developmental character of time.

It is easy to demonstrate that the concept of time has undergone many historical changes. But this is not what is being claimed. What is being asserted is that time has its own genesis and evolution.

How may such an idea be tested? As in the beginning of this chapter, we again turn to Newton.

The phenomena of our concern are the temporalities implicit in the theorems of modern physics. The principle of temporal levels thus bears the same relation to theorems of physics as the theorems themselves bear to laboratory data. Our laboratory data is comprised of all those theorems of science that have something to say about time. To test the principle of temporal levels we will, therefore, examine many of those theorems. We will judge the principle of temporal levels as "accurately or very nearly true," until such time as new scientific theorems are found, tested, and validated, which themselves do not fit the proposed understanding of reality.

Each of the following chapters concerns itself with all the temporalities, but the foci of the chapters shift according to the theorems being considered.

3

Special Relativity Theory

Kinematics and Electrodynamics Tied to Atemporal Motion

The far-reaching consequences for physics of the special theory of relativity follow from the fundamental role that the theory assigns to an atemporal process, the propagation of light. The hierarchical theory of time sees atemporal processes as constituting the primordial substratum of the universe. An examination of the time-related teachings of the special theory of relativity is, therefore, a convenient beginning of a critical assessment of the principle of temporal levels.

3.1 From Absolute Rest to Absolute Motion

In Newtonian kinematics a universal framework of absolute rest and absolute time was assumed to correspond to the true and real nature of the universe.[1] The mental image of absolute rest is one of changeless, motionless existence. Before the age of Copernicus the earth was understood to have defined such a framework. Horses, ships, and stars were said to move because they changed their positions with respect to the earth. The Copernican revolution extended the meaning of the word *rest*. Absolute rest was now attributed to the solar system and changes of position were assigned even to the earth, relative to the new framework. Should someone have given evidence that the sun itself was in motion, absolute rest could still have been sought in Aristotle's fifth or heavenly element, the incorruptible ether. That postulated substance did, in fact, reappear at the end of the eighteenth century as the oscillating medium of light waves.

The speed of absolute rest, a velocity of zero, was taken to be invariant for all observers. Anyone who understood physics could determine, at least in principle, whether or not a particular frame was moving with respect to the frame of absolute rest. The invariance of the zero speed of the absolutely resting system must have appeared to all concerned as an obvious fact of nature. It was no more questionable than the inevitability of death or the fluidity of water at room temperature; it only demonstrated the self-consistent logic of reality.

In 1905 Albert Einstein observed that "the phenomena of electrodynamics as well as mechanics possesses no properties corresponding to the idea of absolute rest." Henceforth absolute rest ceased to be an obvious fact of nature, it did not any longer represent the logic of reality. But there was something in nature that did correspond to the idea of absolute motion. Light, wrote Einstein, "always propagated in empty space with a definite velocity c, which is independent of the state of motion of the emitting body."[2] Special relativity theory demands that we dispense with all references to absolute rest and rebuild our ideas of motion with respect to a new absolute condition, that of constant unrest. Instead of a structure such as the earth, the sun, or the luminiferous ether, the new reference was a process, a very rapid motion. A strikingly novel and unexpectedly fruitful understanding of the physical world followed.

Both in the popular and in the professional literature, discussions of time in special relativity theory reflect the experiential primacy of Newtonian physics. First the idea of a local inertial frame is usually defined. It may be that of a laboratory on earth, or one on a spaceship in uniform translation. Arguments are then developed, using the principles of special relativity theory. Eventually the proper framework of the photon is reached. One enters its umwelt in imagination and notes that in the coordinate system traveling with the photon future, past, or present (separating future and past) can be given no theoretical or empirical meaning. If the same book also contains an introduction to general relativity theory, it is likely to remark somewhere that the universe began in a state that, close to its origins and for a very brief period of time, was made up entirely of light.

It is easier to follow a story time-forward than time-backward. Also, because the universe originated in a burst of light, it should be more appropriate to explore special relativity theory by beginning with the photon than with the proper framework of the galaxy. So it is at the beginning that we begin.

All instants in the life of the photon are simultaneous: it is emitted from the Orion nebula and absorbed in a counter on Mount Palomar at the same time. But no two instants may be closer in time than being simultaneous. It follows that the speed of light, as measured in any of the temporal integrative levels above the atemporal, must be the fastest possible connection between spatially separated events. The limiting nature of c thus follows from the atemporal character of light.

Motion at the speed of light is the oldest kind there is: it was the speed of all objects after the intrinsically chaotic period of creation. It is also the most primitive kind of motion: photons, gravitons and neutrinos remained the surviving ancestors of everything that evolved. The variable characterizing the motion of the photon, its speed c, has retained a unique and invariant relation to all other states of motion that have subsequently become possible. The principle of relativity, a demand for uniformity,[3] is an expression of that invariance.

That no physical meaning can be attached to relative speeds over c appears puzzling because as a species we have no experience of velocities that immense. Had we evolved with feet that carried us walking at $0.99c$, the limiting nature of c would appear to us as obvious. All its experiential consequences would be embedded in our languages and ways of thinking. The existence of the boundary would appear to us as a self-evident fact of nature not in need of special defense or further elucidation. It would be as obvious as the inevitability of death or the fluidity of water at room temperature.

But we could not have so evolved. As is known from special relativity theory, at speeds approaching c the inertial masses of bodies increase without limits. Because of the hierarchical organization of nature, increasingly rapid motion also demands a simplification of the moving structure itself. If the foolhardy experimenter wants to run faster and faster he must first change into a resilient virus, then to a molecular or atomic vapor, and finally to radiation. It is not possible to enter the atemporal physical world without losing the structural and functional complexity responsible for temporal ordering.

Eddington described the speed of light as the fundamental velocity.[4] Reichenbach called it by a felicitous phrase, the first signal.[5] We will think of the propagation of light as one of absolute motion because its role in relativity theory is identical to the one played in Newtonian physics by the concept of absolute rest. In classical kinematics all velocities may be referred to the invariable speed of absolute rest. In relativistic kinematics all velocities are referred to the invariable speed of absolute motion.

Newton's careful handling of the concepts of absolute rest and motion show his awareness that his demands for them are primarily philosophical, and that a reference system of absolute rest cannot be identified in "our region" of the universe. But he believed that an absolutely resting reference system does, nevertheless, exist and that some distant bodies in the remote regions of space might, in fact, represent that system.[6] Thus, Newton's famous definition of absolute motion amounts to a scientific program to be developed by later discoveries and tests. It nevertheless assumed and retained the power of a valid hypothesis because it was possible to deduce from it a system of self-consistent physical laws whose predictions were confirmed by experiment.

According to Newton's first law of motion, every body continues in its state of rest or uniform motion in a right line, unless it is compelled to change that state by forces impressed upon it. In the complete theory of relativity every photon continues in its state of motion as long as it retains its identity. The four-dimensional locus of its motion becomes a right line by definition. The velocity of its motion will be determinable and found uniform by all observers. In its way, relativity theory completes the program implied by Newton's search for a final reference: it provides a theoretically useful and empirically identifiable absolute frame. For objects representing the new reference frame we need not even search "the remote regions of fixed stars, or perhaps far beyond them." They may be found everywhere.

3.2 The Lorentz Transformations

The symmetry of all motion with respect to the atemporal process of light propagation finds its formal expression in the Lorentz transformations of relativity theory. These are rules governing the exchange of time and distance readings among coordinate systems in uniform, linear translation. They satisfy two demands placed upon them by special relativity theory: the first is that they leave invariant the velocity of light; the second is that for small velocities they reduce to the corresponding expressions of Newtonian kinematics.

A geometrical representation of the Lorentz transformations, one of remarkable clarity, may be found in the work of R. W. Brehme.[7] The physical situation is the simplest possible one. Two objects A and B move at a uniform relative speed along a single straight line. To each object there is attached a single spatial coordinate, colinear with its motion, along which its distance from the other object may be measured. Each object carries an identical clock, with the two clocks having been synchronized according to certain straightforward instructions by Einstein. Each object carries a physicist who knows how to determine the time of an event at a distance, using the Lorentz transformations, of which the diagram itself is a representation. Thus it is really we who are examining the goings on, now from the point of view of A, now from the point of view of B.

In the Brehme diagram of figure 3 we see two coordinate systems, A and B, each with a distance and a time axis. They record the distance and time information available to each physicist. To emphasize the formal symmetry of the diagram, all time readings are multiplied by c and plotted with the dimension of length. (We could have divided all length readings by c and plotted them with the dimension of time.) The origin O represents a position $x = O$ at the instant $t = O$ when the two bodies are at the same place at the same time, along their line of motion. We need only add that in relativity theory the world line of an object is the locus of all those points that represent the motion of the object in terms of consecutive coordinate positions of distances and times.

The universe of our two objects is two-dimensional. It has one dimension of length and one of time. An event in this universe is unambiguously identified by a point E in the plane of the paper. The time and distance coordinates of E are the orthogonal projections of E upon the appropriate axes.

The line that bisects the central angles of both the A and the B system is the world line of a light signal. Consider events E_1 and E_2 in the life of a photon that passed the moving bodies at the origin. Inspection will reveal that, because of the symmetry of the diagram, the velocity of such a signal, as seen from either coordinate system, is the same. The requirement for the invariance of c is thus fulfilled.

It may be easily shown that the angle α relates to the velocity of the two systems through the equation

$$\cos \alpha = \sqrt{1 - v^2/c^2}.$$

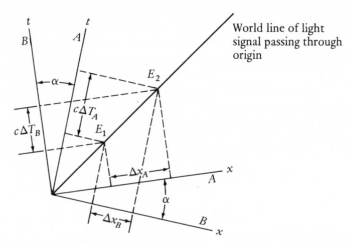

Figure 3. Geometrical representation of the Lorentz transformations, after Brehme. The velocity of light c and the relative velocity of the two systems v are the same when measured in either system. The angle $a = \text{arc cos} \sqrt{1 - v^2/c^2}$ defines a relativistic transformation instruction. For relative velocity c, $a = 90°$ and the space and time coordinates of system A collapse into the world line of light. As the relative velocity begins to decrease we observe the special relativistic creation of distances and times from the atemporal world of light. As the relative velocity approaches zero the relativistic transformation rule approaches the velocity addition law of Newtonian physics, known as the Galilean transformations. Distances and times are seen to be pivoted, geometrically speaking, about the absolute motion of light.

As the relative velocity v becomes small compared with c, the angle α approaches 90°. The A coordinate system then collapses into the world line of light. Since the two time and the two space axes are now orthogonal, two events separated by an amount of time in B will be simultaneous in system A. All finite distances in system B will appear as extensionless points in system A. Furthermore, in system A, distances and times will become identical and hence indistinguishable.

Working the scenario backward—decreasing angle α from its limiting value of 90°—we watch the special relativistic creation of distances and times as they emerge from the atemporal matrix of the universe, represented by the world line of light.

Figures 4 and 5 are coordinate systems in relative motion, the configuration being identical to the one described for figure 3.

Figure 4a is the Brehme diagram of a rod stationary in system *B*. The world lines of the two ends of the rod are plotted orthogonally to the X_B axis. Inspection will reveal that, as measured from system *A*, the rod will be foreshortened by the factor

$$\cos \alpha = \sqrt{1 - v^2/c^2}.$$

Figure 4b is completely symmetrical with figure 4a. It shows that a rod at rest in system *A* will be foreshortened by the same relativistic factor, when measured from system *B*.

Figures 5a and 5b depict two events occurring at the same place but at different times in one or in the other coordinate system. Again, the figures show the geometrical operations corresponding to the Lorentz transformations. Once again, inspection reveals that a period of time, measured between two events at the same place at one system, will be assigned a different value when measured from the other system. Specifically, their temporal separation will be seen as lengthened by the factor

$$\sqrt{1 - v^2/c^2}.$$

The foreshortening of the moving rods is known as the Lorentz-Fitzgerald contraction. It is named after the two men who, independently, suggested it as an explanation for the negative outcome of the celebrated Michelson-Morley experiment. The elongation of the time period is known as *time dilation.*

There are three issues concerning these motional changes to which we will return in section 3.5 and later, but they ought to be recorded here.

First, for the conditions here envisaged, the changes described are completely symmetrical between the two systems.

Second, the changes are not virtual but real.

The third one concerns the kind of differential indication that was imagined in section 1.2. If two clocks in relative linear translation, using light signals for communication, indicate 9,876 and 6,789 time units of separation between the same two events, which one is correct? By the rules of time measurement neither in itself can be said to have measured time and hence be either right or wrong. But if the Lorentz transformations do apply to the kinematics involved and if the relativistic factor can account for the differ-

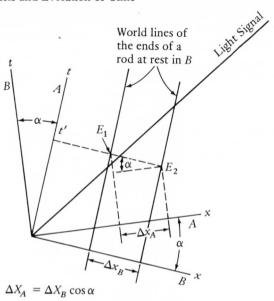

World lines of the ends of a rod at rest in B

Light Signal

$$\Delta X_A = \Delta X_B \cos \alpha$$

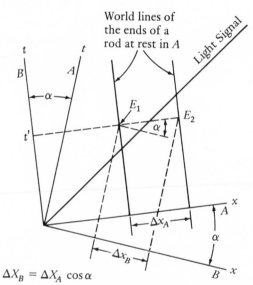

World lines of the ends of a rod at rest in A

Light Signal

$$\Delta X_B = \Delta X_A \cos \alpha$$

Figure 4a, above and 4b, below. A rod stationary in either coordinate system will be found foreshortened by the special relativistic factor $\sqrt{1 - v^2/c^2}$ when measured from the other system, following certain simple rules of measurement given by Einstein. Events E_1 and E_2 stand for "the end of the rod at this point at this time" with respect to either coordinate system.

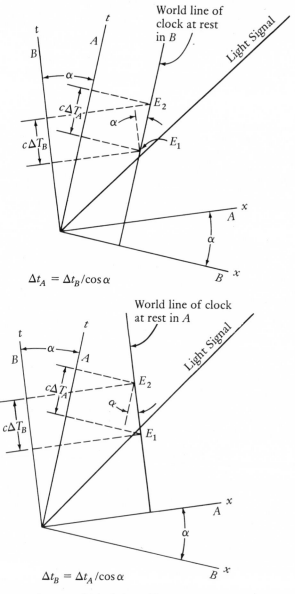

$$\Delta t_A = \Delta t_B / \cos \alpha$$

$$\Delta t_B = \Delta t_A / \cos \alpha$$

Figure 5a above, and 5b, below. Two events occurring at the same place but separated by an interval of time Δt in either coordinate system, when measured from the other coordinate system, will be found to have been dilated or lengthened by the relativistic transformation factor.

ence between the readings, then a satisfactory time measurement has been performed and both clocks are correct.

Time at a distance as determined from local measurements is called coordinate time. The local time itself is called proper time. But we know that to measure time we need two processes. One of these will always be at a distance and in translation, even if the distance is small and the translation velocity is zero. The rules connecting the coordinate and the proper processes—or times, or clocks— are the Lorentz transformations. They are always hidden in a time measurement, even if it is one that ordered the fates of the man, the friend, the cheese, and the dog. Implicit, therefore, in all time keeping is the absolute motion of light. The atemporal substratum is thus beneath all temporal organization.

Assuming now that the universe is homogeneous and infinite in extent, it will be necessary to allow for an infinite number of possible coordinate systems in uniform relative translation, traveling in all possible directions and at all possible velocities. Clock readings taken in any two of these systems will be found symmetrical and mutually transformable by stable rules, fulfilling the demand of the third rule of time measurement. The art of calculating clock readings has thus been mastered, using the motion of light as the absolute and ever-present cosmic reference.

A curious situation has, however, been reached. It is analogous to knowing the exchange rates between any two currencies on earth but knowing nothing whatsoever about the purchasing power of any of the currencies, anywhere.

We know that with the assistance of the special theory of relativity we can define *now* at a distance, provided we know what is to be meant by *now* right here. We can calculate the rate at which time passes elsewhere, provided we know what is meant by the rate of time's flow right here. And we can give meaning to future and past with respect to a distant now, provided we know what we mean by future and past with respect to the now, right here.

3.3 Time and the Minkowski Diagram

Instead of examining how distances and times pivot about the motion of light in a two-dimensional world, let us advance to a three-dimensional world: two spatial and one temporal dimension. We will consider motion in the X-Y plane of figure 6. The third axis is that of time. Instead of a single world line representing the

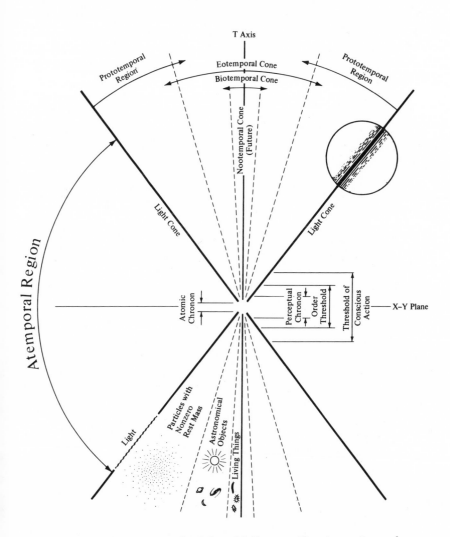

Figure 6. The structured Minkowski diagram. The time axis stands for different temporalities depending on the integrative level of nature in which world lines are being considered. The dashed lines suggest the boundaries between adjacent organizational levels. They do not share the invariant character of the light cone.

pulse of light emitted from a point source in four-dimensional space-time forms a four-dimensional cone. The light cone of the figure is a map of the four-dimensional cone onto a space of two spatial and one temporal dimension.

Motion in the X-Y plane in general will be represented by world lines in the X-Y-T space. The reference epoch is a local simultaneity of two instants: the reading of a clock at $t = O$ and the burst of the light signal. At that reference epoch we make a rapid survey of the world or, rather, we imagine that we do, and plot time units along the T axis and distances in the X-Y plane.

Special relativity theory permits us to imagine any kind of motion in the X-Y plane as real, and hence to take any world line as a possible one, provided the line remains within the light cone and does not double back upon itself. But there are other restrictions that are not at all evident from the unstructured diagram. Nature does not permit us to select entirely freely such objects as would perform the motions represented by otherwise legitimate world lines.

The world lines making up the light cones are called *nul geodesics*. Objects following nul geodesics must be particles with zero restmass, their motions must be atemporal processes. Time stops right along the light cone: that surface and all the regions external to it are atemporal with respect to the here-and-now. Because the dynamics of atemporal processes can supply no temporal information, the timelike nature of the variable plotted along the T-axis must be derived from the umwelt of the surveyor at O and not from anything that happens along the light cone itself.

World lines filling the region of the X-Y-T space close to and just inside the light cone must be those of subatomic and atomic particles. They cannot be those of express trains or spaceships.[8] Imagine that the whole diagram is a representation of a small region of the universe fifty- or one-hundred-thousand years after the big bang: there is nothing around but radiation and particles. The time plotted along the T-axis could not be anything but the statistical, discontinuous, and arrowless time of the prototemporal world.

Leaving the regions of the diagram with the imaginary sign, "For particles only," we come to the region for massive astronomical objects such as galaxies. The world of galaxies as a whole is not included in this region because the universe, as a single object, is not special relativistic in its behavior. The kind of time to be plotted along the T-axis, if it is to be appropriate to the physical reality of the world depicted, must be eotemporality: a continuous time of pure succession, without an arrow.

The world lines of living organisms lie within a very narrow biotemporal cone. As already proposed, and yet to be elaborated in detail in section 7.1, nowness acquires meaning only in terms of the life process. Future and past, as they are usually imagined in the Minkowski diagram, can correspond to reality only if there is a living organism at the space-time point we call the origin.

Finally, the world line of the human observer is a cone collapsed into the T-axis. Only the mature human mind can provide the kind of sharp polarization between future, past, and present which must be assumed before the great significance of the Minkowski diagram can be demonstrated. The region around the origin, in its relation to the human sense of time, is discussed in section 7.4.

It must be concluded that the time axis of the Minkowski diagram stands for different temporalities, as taught by physics itself. Special relativity theory permits the emergence of futurity, pastness, and presentness but it does not disclose the means whereby the potentiality of directed time can change into actuality. It cannot. Thus, the observer at the origin of the structured Minkowski diagram is not an ornament placed there for purposes of instructions. It is there to determine the temporalities above the physical ones, implicit in the way the diagram is drawn and is ordinarily interpreted.

The dashed lines suggest the boundaries between adjacent integrative levels. They do not share the invariant properties of the light cone. They only remind us of the hierarchical organization of the world. Without that understanding it is not possible to reconcile the undirected time of special relativity theory with the directed time of living organisms. Nor is it possible to reconcile the continuous time of special relativity theory with the fragmented time of the atomic world and with the atemporal character of the radiative universe.

3.4 The Atemporal Boundaries of the Special Relativistic World

Having learned about the temporal structuring of reality within the light cone, we may now approach the atemporal boundaries themselves.

It is easy to imagine an object at rest in space but moving, metaphorically speaking, in time. Considering the complete symmetry of space and time in special relativity theory, what would be the complementary condition? It would be an object at rest in time but moving in space. In different words, it would be something unchanging but in ceaseless motion. Photons are such objects. They form the boundaries between world lines entirely within the light cone (known as timelike world lines) and those entirely without (known as spacelike world lines). Eddington wondered long ago what it would be like to encounter a spacelike particle. It would be something, he said, that appeared to exist only for an instant coming out of nowhere, as it were, and disappearing into nowhere.[9]

Let us place a magnifying glass upon the symbol of the light cone in figure 6. As is usually the case when looking through a magnifying glass, what earlier appeared as a sharp line is now seen as a fuzzy and ill-defined region. We see a band within which spacelike and timelike world lines cannot be distinguished and wherein all objects are both things and processes. Away from that region, toward the here-and-now, the world is temporal. Away from the region and further from the here-and-now the world is atemporal.

Judging from our own nootemporal umwelt, the atemporal world is manifest as a spectrum of atemporal intervals called chronons. Reasoning based on the uncertainty principle (sect. 4.1) leads to the conclusion that there exists a lower limit to time and distance below which no meaning can be given to lawful physical process. This limit is around 10^{-23} seconds; we will call it the atomic chronon. It is the lower limit to a continuous spectrum of chronons that together form the atemporal boundary of the special relativistic universe and constitute the absolute elsewhere to the here-and-now. Select a point in the X-Y plane, connect it to the origin, you have the radius of an arbitrary light sphere. Produce the orthogonal projection of the point upon the light cone and the orthogonal projection of that point upon the time axis. The coordinate reading on the time axis is the length of the chronon corresponding to the radial distance in the X-Y plane. When that distance is the radius of the classical electron, we have the atomic chronon. When the distance is the radius of our galaxy we have a galactic chronon; it is about 10^{12} seconds long. General relativity teaches that there exists an upper limit to distance, the value of a variable called the radius of the universe (chap. 6), and hence an upper limit to possible chronons. It is estimated to be 2×10^{17} seconds.

Even without reference to a cosmic chronon, however, we face a dilemma. The atomic chronon appears to us as being atemporal inside, as it were, a 10^{-23} second-long atemporal insert into the flow of time. The galactic chronon appears to us as being atemporal outside, as it were, a 10^{12} second-long temporal insert into the atemporal matrix of the universe. Where is the deflection point? Is there one?

Let someone light a candle at the edge of the galaxy. I can hardly call it a fact unless I have become aware of it. Thursday, just before noon, the signal arrives, having been on its way for 10^{12} seconds. It will still not be part of my reality. The light must first impinge upon my retina and generate a signal, a prototemporal process. Subsequently the disturbance must change into a nerve signal, which is an eotemporal-biotemporal process. Finally it must become manifest in my conscious awareness so that it may become a part of my noetic reality. If I wish to communicate a decision based upon the signal back to the candle, the information must journey back the way it came.

The width of the mental present may be, for the sake of roundness, 200 milliseconds, negligible compared to 10^{12} seconds. The galactic chronon feels atemporal outside because it is longer than the mental present. This is not the case for the atomic chronon or even for a cell chronon of, perhaps, 10^{-7} or 10^{-8} seconds. These periods are much shorter than the mental present. We cannot receive a signal from the beginning of the cell chronon and return a decision to reach it before it ends. It feels as if it were atemporal inside. What essentially happens is that the speeds of communication along our nerve paths, necessary to maintain the simultaneities that keep us alive, are much below that of light.

An inflection point in the spectrum of chronons, a point where they change from being "inside" to being "outside" time exists for us, but not for inorganic nature.

The hierarchical temporalities of the special relativistic world tell us that our biotemporal and noetic universes anchored to our here-and-nows float, metaphorically speaking, in an atemporal sea. The metaphor suggests an identification of the substance of the atemporal world with the classical element of ether. But that classical notion, both in science and in philosophy, is associated with the idea of absolute rest whereas the atemporal world revealed by modern physics is a universe of ceaseless motion.

3.5 Certain Consequences of Nowlessness

In section 3.1 we learned about two symmetries of nature. One was of motional changes mutually observed in two coordinate systems in uniform relative translation. Each observer measured the same amount of time dilation and motional contraction in the other system. The other complete symmetry was between moving rods and moving clocks. Traveling clocks decrease their rates and traveling rods shrink their lengths by just the right amount to keep the relative velocity of the two systems and the velocity of light invariant. This second, dynamic condition is often stated by saying that distances and times play similar roles in the space-time of special relativity theory.

The motional changes of special relativity theory must be taken as real in the following sense. Only if they are assumed to represent actual physical changes do we get a self-consistent description of nature. Physics does admit auxiliary variables which some writers like to call fictitious, such as Coriolis and centrifugal forces. No one doubts the existence of gradient winds attributed to these forces. The only debate is whether the corresponding physics ought to be written in a nonrotating or in a co-rotating frame. The motional changes of special relativity theory are not of this kind because the theory governs the space-time structure of the physical world and we have only one of those. But to make sure that we understand the motional changes of the special theory, let a clock and a rod be sent on a round trip at relativistic speeds. Through measurements during the trip and later, after their return home, we expect to observe the reality of those changes.

A round trip by the same object will necessarily involve nonuniform motion which, it is sometimes said, cannot be handled by special relativity theory. It has been suggested, therefore, that the accelerating phases be made to take up a negligibly small portion of the total travel time. This, unfortunately, is an unacceptable condition because significant physical changes often take place in negligibly short times. A better solution is to realize that proper acceleration may be very well handled in special relativity theory.[10] Then, on Einstein's authority we may assume that clock rates depend only on instantaneous velocities and not upon the rates at which the velocity changes. Or one may get around the problem by using not one but two or more traveling clocks, each in uniform translation and all mutually synchronized according to self-consistent plans.[11]

Let the test be completed by any acceptable means and the clock and rod readings compared. We will find ourselves with two problems.

The first one was already mentioned. Why did the traveling clock or combination of traveling clocks register 6,789 units and the stay-home clock, 9,876 units and not vice versa? An answer to this question will be given in section 6.4.

The second problem is somewhat surprising.

Upon their return the traveling clock will again tick at the rate of the stay-home clock and the rod will again be the length of the stay-home rod. But an asymmetry will be manifest between the two devices. The rod will have bounced back to its original length of 1,000 millimeters from its mid-journey length of 687.4241 millimeters; it will show no effects of the trip whatsoever. It will not have been crushed, deformed, or mutilated. But the digital display of the clock will keep on showing 3,087 less ticks than the other one does, a permanent reminder of the trip it has taken. Or, if you wish, the extra 3,087 ticks registered and remaining upon the stay-home clock will tell us about the vicissitudes of having been at rest.

I have insisted that motional changes must be taken as real and also that time dilation and length contraction are symmetrical phenomena. It seems now that the phenomena are, after all, not symmetrical. Or, perhaps, that the time dilation is somehow "more real" than the contraction. What is going on?

It will be remembered that eotemporality is a time of pure succession. The notions of "passing time" or "aging" cannot be applied to eotemporal processes, not even metaphorically. In any case, time measurement in itself does not demand the concept of passing time or aging. It only demands two processes connecting two events, two sets of numbers and appropriate transformation equations. If we had two clocks side by side and one of them indicated a larger number than the other, would we say that they existed for different lengths of time? No. Likewise, all the two clocks did in our experiment was to supply the ticks, we supplied the transformation equations and we measured time.

The physical universe can determine only simultaneities of chance; it has no present, future, and past; no arrow. If we did convince ourselves of the correctness of motional changes, instant by instant, we have done everything necessary to demonstrate the symmetry between distances and times and the reality of the changes. The cumulative count read on the display was imported into the eotemporal world from the biotemporal and higher organizational levels. If time does not have a direction it cannot be cumulatively counted—there is no way of knowing whether the next unit is to be added or subtracted.

Instead of two clocks and two rods we could have used two radioactive meter rods. After their reunion the two will again have the same length but the stay-home one will have accumulated more decay products than the wayward one did. The situation is analogous to the prior test. Instant-by-instant confirmation of special relativistic rates and lengths is sufficient to demonstrate the reality and symmetry of the phenomena.*

Next, let us employ two mice. The traveling mouse will return home 3,087 units of time younger than her less adventurous sister. The introduction of the life process broke the eotemporal symmetry of the physical world, but only as far as the organisms are concerned. The physical world allows asymmetrical aging—it also allows Holy Wars—but in the absence of a physiological present, it cannot give any meaning to the aging process itself. It will be useful to remember that one cannot even count hours, days, or years without reference to directed time whose direction could not be determined from cyclic changes.

Instead of sending ticking clocks or running mice to far away places, we may examine the effective internal nuclear frequencies of iron nuclei, moving at different rates. This was, in fact, accomplished by comparing very precise frequencies of gamma rays from iron nuclei in crystals at different temperatures.[12] The nuclei of the warmer sample were running around faster, they were the traveling processes. Their times were found dilated by the amount predicted from relativity theory. In one of these tests thermal vibrations produced velocities of the order of $10^{-6}c$ (500 meters/sec) and accelerations of the order of $10^{16}g$. The experimental results showed that time readings were independent of acceleration. Within the range indicated, it confirmed the validity of Einstein's conjecture that time rates depend only on instantaneous velocities.

Taylor and Wheeler writing about the same experiments exclaimed, "fewer vibrations; fewer clock ticks; fewer birthdays; more

---

* Unlike the clock readings, the accumulation of decay products brings in the issue of reversibility and irreversibility. See section 5.4.

youthful!"[13] But youthfulness, even metaphorically, is a misleading concept in the context of this discussion. We might paraphrase an earlier quote, "Ask any iron atom whether it is young or old, and it will laugh at the question."

The consequences of the nowlessness of the physical world are sometimes strange. But that is a statement about our understanding of the world, not about the world itself.

3.6 About Absent Arrowheads

What is meant by the claim that classical kinematics is insensitive to the direction of time?

Drop a ball from the Leaning Tower of Pisa, straight down. Photograph its fall on a motion picture, and develop the film. Before copying it, block out the background. Project it in the forward direction and question your audience. Did they see (a) a ball falling from rest to rest, photographed, projected forward in time, or (b) a ball thrown upward with a certain initial velocity, photographed from rest to rest and projected in a backward temporal direction? You may also perform the conjugate experiment by throwing the ball upward and photographing it. The viewers will not be able to answer your questions because the kinematics of fall and rise are completely symmetrical with respect to time. It is true that balls are more likely to be dropped from high towers than thrown upward to their heights. But the frequency of natural occurrences of one, as against the other, is of no significance.

The symbol t in the equations of motion that govern the rise and fall of balls stands for pure succession. Neglecting thermodynamics, even the dynamics of the two processes are symmetrical: the taut spring that jettisons the ball upward will be depressed back to its initial state by the falling ball. The tightness of the spring may be seen either as a cause or as an effect depending on the direction of time, obtained from external sources. This is an example of two-way, deterministic causation.

One may think up many other examples. But for causality in physics one need not appeal to special causes and effects. One begins with a mathematical relationship of certain variables so that, from the data available for epoch t_2 it is possible to calculate the state of the system at epoch t_1. The temporal sequence of the epochs does not matter; cause and effect are interchangeable.

The prototype of a successful deterministic, two-way, or eotemporal theory, is the Newtonian theory of mass points and whatever one may construct using mass points: bodies made of molecules, a universe made of galaxies. Maxwell's equations treat macroscopic models of charges and fields the same way Newtonian physics treats masses with gravitational fields. Therefore Maxwell's equations possess the same eotemporal symmetry as do those of Newtonian mechanics.

It has often been asserted that except for a very few special cases, all physics is of this two-way nature. The cases judged unusual have been regularly singled out for attention because they are believed to demonstrate the origins of time's arrow in physical process. We will examine the most frequently given examples of one-wayness in physics, short of getting into thermodynamics.

Milne has noted that a swarm of particles contained within a final volume, each moving along a straight line, if originally contracting, will become irreversibly expanding. In contrast, a swarm of expanding particles will not, on its own accord, become a contracting one. Whitrow described Milne's model as "a remarkably simple but fascinating example of irreversibility. . . ."[14] He added that it is sufficient to consider only two particles. If they are initially approaching they will eventually move apart; if they are moving apart, they will never again approach each other.

Let a laboratory be placed on one of the particles and let us sit down on its bench, looking out the window. If a particle is approaching from the left it will disappear on the right and vice versa. These facts will not help us with the direction of time. If a particle is already moving away it will keep on moving away. But how does one know that it is moving away rather than coming closer? By an appeal to the sense of time. If the sense of time is removed it is not possible to determine the direction of motion. Whenever a velocity vector is drawn, reference is implicitly made to the direction of time. If the origin of the vector is point P_1 and the vector is directed toward P_2, the symbol says that the object was, is, or will be at P_1 before it reached or will reach P_2. Without an arrow of time, obtained from sources other than the swarm of particles itself, the behavior of the group cannot be used to demonstrate the direction of time.

Central wave propagation is a name given to another process judged to be irreversible and, hence, a demonstration of the one-wayness of time in the physical world. Attention is drawn to the fact that waves diverging from a central source are frequently observed in nature, whereas we never observe waves that converge upon a center.

Imagine a quiet pond—perhaps not too far from the man with the cheese and the dog—and drop a stone in its center. The stone will sink and leave in its wake a ring of waves propagating outward. A film taken of the propagating waves and projected backward will show a ring of waves converging upon the center and there so interfering as to produce a quiet surface. To create such converging waves it would be necessary to coordinate the phases of a large number of oscillators, placed along the circumference of the pond. This may be done only by providing a central control, such as a signal sent from the center. Except for central control, in the words of Karl Popper, "the contracting wave, though not in itself physically impossible, would nevertheless have the character of a physical miracle." In the absence of a prepared plan, "although theoretically possible, spontaneous contracting waves do not, in fact, occur. . . ." They are "causally, and therefore physically, impossible."[15]

The crucial specification here is the spontaneity of the process opposing the expanding waves. I wish to argue that there exist natural processes representing the kind of organization that Popper described as impossible and that they do not occur in response to prepared plans. At the end of the reasoning we will refer back to the issue of central wave propagation.

Converging waves on a pond could be produced by dropping a ring on its periphery. There are two problems with this scheme, however. First, it clearly involves plans. Second, the gadgetry is not easily available. The reason for its absence from catalogues is that neither man nor beast nor stone nor vapor seems to have much use for it. But whenever nature in its evolutionary passage does have use for processes exactly analogous to the converging waves, those processes are clearly manifest.

It is known, for instance, that operating through physical and chemical processes, living organisms can concentrate an element as much as a million times the average abundance of the element in nature. The organism functions in complete analogy to the converging waves on the pond. When it dies, its molecules become the diverging waves or diffusing elements of life.

The question to decide is whether or not the concentration of elements in living matter is spontaneous. Is the formation of amino acids, purines, and DNA molecules spontaneous? Is the life process itself? If the relative likelihood of finding turtles in the universe is a measure of spontaneity then life can only be regarded as a result of design. The millionfold concentration of elements in living organisms becomes one of the many ways in which the design may be illustrated.

I prefer to opt for an alternate view, for maintaining that biogenesis, organic evolution, and even the mental and social life of man ought to be explainable within the natural, unplanned order of things. Their rarity in the universe does not, by that account, make them either impossible or in need of a designer. I hasten to add that the absence of a designer does not make these phenomena any less awe-inspiring. In any case, unless someone can provide a principle that explicitly prohibits such phenomena as converging waves, the statistical preponderance of expanding waves cannot illustrate the one-wayness of time in the physical world. The immensely greater likelihood of seeing objects dropped from the Tower of Pisa as against observing objects jettisoned upward to it does not, on that account, remove the symmetry of classical kinematics.

The electromagnetic version of central wave propagation has been widely discussed under the heading of advanced and retarded waves. Maxwell's celebrated electromagnetic equations have two solutions for the potentials of electromagnetic radiation fields. These solutions express quantitatively the magnitude of a local electromagnetic disturbance due to a distant source. They allow for two causes of locally observed effects: (1) a distant source active in the past and (2) a distant source active in the future. The solution that describes the effects of the distant source active in the past is called the retarded potential solution. The one describing the effects of a distant source active in the future is called the advanced potential solution. The equations of propagation are completely symmetrical with respect to time and, hence, one would expect to find examples of signals arriving here and now, having been emitted in the past and also signals arriving here and now, yet to be emitted in the future.

But there is no record of anyone's ever having received a signal from the future; no one has ever heard next morning's broadcast the night before. There seems to exist, therefore, an electromagnetic arrow, an intrinsic property of the electromagnetic world, forcing us to select the retarded potential solution of Maxwell's equations as the only one corresponding to fact, and neglect the advanced potential solution as nonphysical.

The search for the roots of nature's preference for the retarded waves has been going on since the first decade of this century when W. Ritz and A. Einstein debated the issue.[16] Ritz maintained that the preference is due to the laws controlling the emission, while Einstein held that it derives entirely from statistical principles. Whereas for the retarded solution it is sufficient to know the generating process of the macroscopic source, for the advanced solution it would be necessary to specify fully all the microscopic details of all the boundary conditions.

Since the Einstein-Ritz exchange many ideas have been proposed and some even abandoned, in a continuous effort to explain the physical reasons of the observed asymmetry.

The situation is only partly analogous to central wave propagation. There both the diverging and converging waves proceed along the biotemporal arrow from past, to present, to future. This is not true for the advanced potential solution to Maxwell's equations, for that solution describes a signal that is working its way from the future to the present.

Imagine a vast sphere of atomic oscillators built in the depth of space with us at its center. They are so wired and triggered that an inward moving wavefront collectively generated by them will converge upon a dipole antenna on the laboratory table, there to give a single coherent signal.

How could we know that a signal so received came from the future and not from the past?

Because the radiation came from the future, it ought to be possible to trace its history back to its origins by going opposite to it, from present to the future. First, therefore, we expect to observe that the local antenna begins to oscillate. It received the signal. Then, by instruments placed at increasing distances from the antenna the passing of the advanced wave will be detected. A day after the signal was received monitors, placed one light-day away, will record the passing of the wavefront. Monitors located two light-days away will do likewise in two days. If the source of the advanced wave will emit its radiation Friday noon and the signal was received

Monday noon, then it should be possible to meet the approaching waves on their way to the laboratory come Tuesday, Wednesday, and Thursday. We conclude that, on this account, the two solutions to Maxwell's equations are empirically indistinguishable.

Let us consider the details step by step.

The antenna itself has no way of telling whether the signal exciting it came from the future and a million miles away or from the past and an oscillator next door. Its properties as a receiver and as a transmitter are symmetrical, its umwelt is eotemporal.

Let us include the receivers at different distances with the antenna. The two waves are still empirically indistinguishable but a temporal direction has been smuggled in. As do all directed processes in our umwelt, the two waves both progress from present to future: first an oscillation here, then signals at increasing distances. If we were to position ourselves along one of the distant receivers, we should then be able to tell whether a wave is coming from the right (the direction to the distant oscillators, emitting in the future) or from the left (the direction of our radio transmitter, emitting in the past). But as we have learned, "coming from" hides a direction of time. The umwelt of the retarded and advanced waves, observed en route, is still eotemporal.

The separation of the advanced from the retarded waves, in terms of signals originating in the future or in the past becomes possible only with respect to our physiological present. The device that imposes just the right conditions upon the antenna so that it becomes an instrument exosomatic to our biological and mental functions, is known by the name radio transmitter. The physical world does not have a preference for the retarded wave solution; it has no means by which to distinguish it from the advanced wave solution. The preference is a phenomenon of our biotemporal and noetic umwelts.

All this is implicit in the use of the Minkowski diagram. One talks about future and past light cones. In reality, they are diverging and converging light spheres. The distinction between them is made possible by the symmetry-breaking functions of the life process.

By replacing the Parmenedian framework of absolute rest by the Heraclitean framework of ceaseless motion, special relativity theory has created a new kinematics and dynamics of impressive generality and power. The change amounts to the rewriting of physics from the atemporal integrative level and upward. This chapter has described the emergence of distances and times from the atemporal substratum of the universe, according to the principles of special relativity theory and the principle of temporal levels. It has explored the special relativistic boundaries of the temporal universe and noted some of the peculiarities of eotemporal change.

4

Quantum Theory

The Rules of the Prototemporal World

Soon after the universe left its purely atemporal, chaotic state with a big bang, the era of particle creation commenced. After another 10^{12}–10^{13} seconds the world settled down to enjoy its permanent complement of particles. That new integrative level, the prototemporal one, has been with us since those early instants, having come about even before the galaxies began to form, perhaps 10^{16} or 10^{17} seconds after the cosmic singularity. In the evolutionary view of nature espoused by the hierarchical theory of time, the laws of the prototemporal world must, therefore, be regarded as developmentally and hence ontologically prior to the laws that govern the behavior of the eotemporal world. But historically it was the physics of the eotemporal world that was discovered first because it dealt with human-sized structures and processes.

Special relativity theory must be taken into account in the physics of the particulate world. Some quantum mechanical effects are special relativistic in their origins. But special relativity theory does not disclose the most peculiar features of the prototemporal universe: duality, indeterminism, complementarity, and the probabilistic character of prototemporal law.

4.1 Duality, Indeterminism, and Complementarity

Quantum mechanics teaches that elementary forms of matter may be given two different but equally valid descriptions. One comes from classical optics. It speaks about waves and uses such variables as are appropriate for wave motion: wavelength, phase, frequency, and amplitude. The other comes from classical mechanics and uses variables appropriate for the motion of massive bodies: position, linear momentum, angular momentum, mass, time span.

That something is masslike, such as an orange, or wavelike, such as the sound of a song, is not a very shocking discovery. The two kinds of descriptions exist side by side precisely because they were found useful. New and surprising in quantum theory is that within its domain, (1) both descriptions may be applied to all objects, (2) may be applied at will to suit theoretical and experimental needs, but (3) cannot be applied to the same object at the same time. Whenever an experiment puts a question to nature phrased in terms of one or the other mode of description, nature will answer using the mode in which the question was asked. This peculiar, double characteristic of prototemporal objects is known as *duality*.

Charles Sanders Peirce, physicist, mathematician, and the founder of pragmatism, defined sign as "something that stands to somebody for something in some respect or capacity."[1] The wave and particle representation of the same species of objects amounts to attaching different signs to the same kind of object.

The use of signs to represent physical variables or processes is not at all new. The formula, $v = d/t$, is a combination of three signs. Their relationship corresponds to those among the experimental variables of speed, distance, and time. But whereas in the macroscopic domain we may allow ourselves to directly identify the sign with different aspects of our experience, in the atomic world such direct identification is not possible. As with velocities near c, or with cosmic distances, so with atomic particles: we have no direct experiential knowledge of them. We must learn about them through mathematical physics.

Quantum theory is mathematical physics in its pure form. It seeks mathematical symbols whose behavior with respect to other mathematical symbols describes the behavior of physical observables with respect to other physical observables and thus reveals their relationship. Therefore the semiotic character of quantum theory is strikingly evident, whereas the semiotic nature of macroscopic physics is mostly hidden by convention. Prototemporal and eotemporal descriptions do, however, meet as they must. As the number of particles whose quantum mechanical descriptions are being considered increases, quantum descriptions become irrelevant and give way to the classical and relativistic physics of macroscopic objects.

In the macroscopic world, although a body may be undergoing change, it is most readily thought of as a structure. It is something primarily spacelike. A wave, although it may have a spatial structure, is primarily associated with the idea of change. It has a very definite timelike quality. But the wave of quantum theory is not timelike, it need not report about change; a particle in quantum theory is not a superminiature orange, it is not a spatial structure. The distinctness of spacelike and timelike features of reality vanishes in the microscopic world. It is the absence of the distinction that expresses itself in the principle of duality. Either the process-like or the structurelike features of a quantum-mechanical object may constitute a complete description thereof, except that they cannot be used at the same time.

What quantum mechanics does is to separate the single nature of a species of objects into two of its aspects because those aspects are familiar to us from human-sized experience. The wave-particle duality of quantum-mechanical objects thus demonstrates the yet incomplete separation of distances from times along the most primitive temporal organizational level of nature. Duality is, therefore, a witness to the emergence of distances and times from the atemporal substratum of the universe. We have already observed this act of creation through the macroscopic representation afforded by special relativity theory. Quantum mechanics reports about the same developmental facts in terms appropriate for the particulate universe.

It is instructive to represent the physical situation in terms of logics. In the prototemporal umwelt an elementary object is a particle *and* a wave, simultaneously. Described in terms of distances and times (distinct in the eotemporal world), the same elementary object is *either* a particle *or* a wave. Because the hierarchically nested organization of nature, eotemporal laws (including eotemporal logic) subsume the laws (and logic) of the prototemporal world. It follows that, in the eotemporal world, an elementary (prototemporal) object is a particle and a wave as well as either a particle or a wave.

Historically, we have had the eotemporal world of mechanics describing the behavior of massive matter side by side with the wave optics of atemporal light. With Newton and Huygens came the maturation of the corpuscular and wave theories of light. Maxwell demonstrated the electromagnetic character of light and thereby made possible the development of special relativity theory. In 1923 de Broglie extended the principle of duality, valid until then only for light, to the theory of matter.[2] His broad and stunning generalization, subsequently confirmed by experiment, permits us the tracing of the evolutionary development of space and time, as well as matter, all the way back to the atemporal substratum of light. Before being able to do so, however, it will be necessary to become acquainted with a few peculiarly quantum-mechanical modes of representation.

In quantum mechanics it has been found useful to employ the concepts of wave (or phase) and group velocities, known from the study of wave motion. Wave velocities are the speeds at which any arbitrary phase of a wave travels. Group velocity is the speed at which a modulation envelope travels. Such envelopes may be generated by linearly superposing waves propagating at the same speed but having different frequencies. The envelope, called the wave packet, is the quantum-mechanical representation of what is a particle in the eotemporal world. But unlike the concept, *particle,* a word suggesting a permanent structure, the concepts—the sign— *wave packet* suggests fluidity. In quantum theory motion itself appears as a consequence of interference among wave trains extending, theoretically, over all space. The wave packet may melt away, dissolve into the spectrum of waves that made it up. This feat has no analogue in the macroscopic domain.

In the de Broglie model of matter, the group velocity corresponds to the velocity of the object with respect to the observer. De Broglie's equation, $\lambda = h/mv$, combines variables appropriate for particles (mass and velocity) and for waves (wavelength). h is a constant of nature, introduced into physics by Max Planck. From this formula we can calculate the matter wavelengths of moving electrons, molecules, and running rabbits. In all cases the wavelength depends on mv, the momentum of the object. For photons $m = 0$, but their momentum, $p = mv$, is finite. Using a relativistic relation between energy and momentum, $E = pc$, the de Broglie wavelength for photons becomes $\lambda = hc/E$ and the matter waves of de Broglie become the electromagnetic waves of radiation. Because phase velocity u and group (particle) velocity v can be shown to relate through $uv = c^2$, the phase and group velocities of the photon coincide; they are both c, the absolute velocity of light.

With de Broglie's findings safely in mental storage, let us return to considering the atemporal boundaries of the temporal world.

It was stressed in section 2.3 that atemporality may be asso-

ciated with the idea of chaos, the state of a world made up of incoherent atemporal processes, but not with the notion of non-
existence. To integrate the atemporal chaos with our understanding
of the history of the universe we will say that, as a universal state
of the world, it preceded Creation (see, sec. 6.4). We thus imagine
biotemporal ordering where there can be none. This way of talking
is warranted by the features of our noetic umwelt. We are able to
give descriptions only in terms of space and time. It is possible to
employ symbols, but they also eventually lead to descriptions in
terms of distances and times; in fact, that is their purpose. Discussing the dynamics of Creation in tenseless language is awkward and
would be just as inaccurate as speaking about chaos having preceded Creation. Those events enter our noetic world of symbolic
transformations as having been in the past and, by the generalized
umwelt principle, their pastness is a feature of noetic reality.

The pre-Creation atemporal chaos comprised electromagnetic
waves of many frequencies. When waves of different frequencies
propagate in a conventional medium, such as water, they will
propagate at different speeds. Water is said to be a dispersive
medium. But all waves of the pre–big bang universe had to travel
at the same speed. The medium of which light is a ripple, unlike
water, is nondispersive. At the instant of Creation the universe
became a dispersive medium making it possible and necessary to
distinguish between group and phase velocities for matter waves.
As speeds began to drop below the absolute velocity of light, objects of finite restmass precipitated and hence separated from the
atemporal chaos, together with distances and times. Group velocities became the measure of particle speeds while phase velocities
by the $uv = c^2$ relation came to exceed the speed of light.

It will now be postulated that from the atemporal integrative
level of the universe may emerge all such processes and structures
as satisfy the boundary conditions imposed upon them by the
temporal universe. In reverse, the atemporal integrative level may
absorb any and all things and processes that can approach that
boundary, without limits. We may imagine, for instance, the phase
waves of matter enter the atemporal integrative level of nature,
there to await their usefulness for explaining various theoretical
and empirical issues of physics.

An object can come to rest next to an observer only if it is an
aggregate of particles held together by internal forces, making up

a macroscopic body. The group velocity of the object then becomes zero; the phase velocity of the associated matter waves becomes infinite. Speeds faster than light are not prohibited by special relativity theory provided no information is propagated at those speeds. In this case information does not travel anywhere at all because the body is at rest. Matter waves at supraluminal speeds represent, therefore, not information available or action at a distance but only potentialities hidden in the atemporal integrative level. Particles, for instance, are such potentialities.

The modern theory of fields explains interactions among particles as exchanges of energy packets, conceived of as the carriers of force fields. The range of these force fields, the distances across which they can exert influence, is inversely proportional to the energy-equivalent of the restmass of the corresponding messenger particle. When the restmass energy is zero, such as for photons, the range is infinite. It is thus that photons and gravitons may be carriers of electromagnetic and gravitational interaction at cosmological distances. As the early universe cooled, nonzero restmasses precipitated, and force fields with ranges of finite distances became possible. The new conditions allowed the trapping and structuring of energy in the form of chemical elements, accompanying the emergence of space and time from the atemporal chaos.

In celestial mechanics one can compute the future fate of a planet from the instantaneous values of its motional and positional variables. The variables found useful in atomic physics are momentum, energy, position, and time. In order to predict the future history of a particle one would want to have accurate, simultaneous values for these variables. But they cannot be obtained. There exists a conspiracy among experimental parameters preventing us from making simultaneous determinations to any desired precision of such pairs of variables as would be necessary for accurate predictions.

Let $\Delta p, \Delta x, \Delta E,$ and Δt be the experimental spreads in the measured values of momentum, position, energy, and time of a moving particle. Since Heisenberg it has been known that these variables, when they pertain to measurements of elementary objects, correlate through two basic equations: $\Delta p \ \Delta x \geqq h/2\pi$ and $\Delta E \ \Delta t \geqq h/2\pi$. There is no limitation on increase of measuremental accuracy (decrease of spread Δ) of any of the variables. But as the spread of one variable decreases, the spread (uncertainty) in the data of the conjugate parameter will increase. The two sets of conjugates are, however, independent. It is possible, for instance, at least in principle, to determine the position of a particle at a precise instant to any desired accuracy. But this will not help in the task of prediction because the associated momenta and energies will be widely spread.

The "conspiracy" prevents us from making anything but statistical predictions about the future behavior of elementary objects because present positional and motional parameters may only be known statistically. The generalized umwelt principle compels us to regard the collective, statistical character of position, time, energy, and momentum in the prototemporal world as a report about the real nature of that world.

It can be shown through a simple thought experiment that the Heisenberg uncertainty derives from the wave-particle duality of quantum theory.[3] But the wave-particle duality demonstrates the incomplete separation of space and time in the prototemporal world. It follows that the uncertainty principle of Heisenberg is itself a corollary of that incomplete separation.

Closely related to the wave-particle duality and the uncertainty principle of quantum theory is the principle of complementarity, first enunciated by Niels Bohr in 1927.[4] According to this principle it is impossible to establish with a single experiment that a particular elementary object possesses both wavelike and particlelike properties. Complementarity as an empirical fact has been demonstrated to hold and there are many fine discussions about the epistemological significance of the principle. But the physical causes responsible for complementarity have not been sufficiently explored.

It is well known that in the world of particles, unlike in the macroscopic world, the very act of measurement tends to demolish the value of the variable measured. But we also learned that in its own umwelt, an elementary object is both wavelike and particlelike at the same time. If I ask an experimental question in the wave language and thereby demolish the wavelike variable in return for an answer, I have also demolished the particlelike variable, and vice versa. The object itself cannot distinguish between wavelike and particlelike questions.

Duality, indeterminism, and complementarity may thus be seen as different manifestations of the incomplete separation between the identities of distances and times in that primitive integrative level we have called prototemporal.

4.2 Probabilistic Law and Quantization

The principle of complementarity prohibits any experimental demonstration of a single object having both wavelike and particlelike properties. It could be claimed, therefore, that the principle of complementarity casts doubt on the principle of duality. Perhaps there are particlelike electrons and wavelike electrons and they sort themselves out depending upon the experiment.

This challenge can be answered by referring to yet another of the curious properties of the prototemporal universe: the complete identity of all particles belonging in the same species. There is no need to regret the loss of an electron to a wavelike test. Another electron will come around, exactly identical to the first one, and may be subjected to a particlelike test.

It is interesting to inquire into the reasons why similar elementary objects are also identical.

The atemporal universe can only be imagined as being without law or connectedness: an intrinsically uncorrelated and complete chaos. Quite in contrast, the connectedness of the eotemporal world is one of two-way, deterministic causation. The prototemporal world is the intermediate evolutionary step between the atemporal and the eotemporal. Probabilistic causation seems to be the necessary step between no connectedness and deterministic causation.

Probabilistic laws can be applied to a set of objects only if the members of the set are distinct and hence countable, but also indistinguishable, at least in that respect of their behavior to which the probabilistic law is to apply. If one were to seek the features of an integrative level between the chaotic atemporal and the deterministic eotemporal ones, a probabilistic universe of indistinguishable objects would suggest itself as the right choice. Quantization (distinctness) and probabilistic law may thus be seen as emerging together as the necessary evolutionary features of a world between chaos and continuity.

The nature of probabilistic law is puzzling.[5] We will discuss one particular law of probabilistic aggregates, widely used in statistical mechanics.

Select a quantity pertaining to a single dynamic system of distinct but indistinguishable members. Measure this quantity by ensemble average and by time average. An ensemble average describes the simultaneous states of many identical systems. A time average is obtained from consecutive states of any one single system. According to a theorem introduced by Ludwig Boltzmann in 1871, later elaborated by J. Willard Gibbs and known as the ergodic hypothesis, these two averages should agree.

The agreement between ensemble and time averages has been so excellent for a wide variety of systems that the ergodic theorem is considered valid in spite of the absence of rigorous justification from first principles. We are left, therefore, with the following question. What basic mechanism provides for the identical behavior of several systems spread out in space and surveyed at an instant, and a single process at one point but spread out in time? What in the nature of the physical world permits the free exchange between spatial and temporal distributions?

The answer suggests itself. A collective of indistinguishable elements determines a prototemporal umwelt and in a prototemporal world, space and time are, for certain purposes, interchangeable.

Quantization as an aspect of probabilistic worlds is not a function of size but of degrees of organization or complexity. Consider, for instance, the behavior of superfluid helium, which is the state of helium below about 2.2°K. When an ordinary vessel is filled with an ordinary liquid, such as a Newtonian pail with water, and is rotated about its axis of symmetry, the liquid will smoothly follow the pail. First it will lag behind but eventually will rotate with the angular velocity of the pail. Superfluid helium refuses to do so.[6] When placed in a bucket (2 mm. across) which is then rotated, the liquid first remains at rest. When the bucket reaches a certain critical angular velocity the liquid develops a flow along a single vertex, determined by quantum-mechanical rules. New vortices are added as the rate of rotation of the bucket reaches and passes subsequent critical speeds. Although the pail of helium is immense compared to the dimensions of the helium atom, it is still a quantum-mechanical system.

The results of this experiment encourage the following conjecture. Although prototemporality as a canonical form of time first appears in the dynamics of elementary particles, it should also be appropriate for the dynamics of all collectives made of indistinguishable objects.

4.3 Experiments and Thought Experiments

There are a number of experiments, thought experiments, and experimentlike questions pertaining to the atomic world which have caught the interest of practically everyone who has ever considered them. I will try to demonstrate that the principle of temporal levels introduces a substantial economy of thought into the issues involved.

The double-slit experiment   Figure 7 is the top schematic view of the kind of apparatus that has been used in the study of optical interference and diffraction.

It depicts an opaque shield with two parallel slits (directed into the plane of the paper) and a fluorescent surface or photographic plate, serving as a projector screen. A source on the left, not shown, generates a light beam or an electron beam that passes through the slits so as to form images of the slits on the screen. It has been known from Young's interference experiments with light in the nineteenth century that, for properly selected experimental conditions the image will not be a simple projection of the two slits but an interference and diffraction pattern. Such a pattern is shown on Screen 1. A simple argument from geometrical optics shows that light waves will cancel at all those points on the screen for which the respective distances to slits A and B differ by an odd number of half wavelengths.

The experiment with which we are concerned studies the interference of matter waves associated with the electron. The source, whose intensity may be controlled, generates not photons but a collimated beam of electrons. The electrons then pass through the slits to form an image of the slits on the screen.[7]

Because of their particulate nature—they weigh 10^{-27} grams each—electrons would be expected to behave as particles: diffract along the edge of a narrow slit and distribute themselves upon a screen, according to a bell-shaped intensity pattern. The pattern itself may be calculated from the theory of diffraction in geometrical optics. This, in fact, is the case. Cover up slit B and, on Screen 2, obtain pattern A', an image of slit A. Cover up slit A and obtain the curve B'. One would expect that with both slits open the combined electron distribution would be the sum of curves A' and B'. But that is not the case.

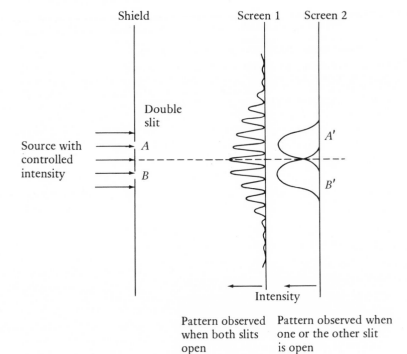

Figure 7. Schematic representation of apparatus used in optical as well as in electron diffraction and interference experiments.

The pattern obtained with both slits open is the kind of intensity distribution one associates with the interference of waves, well known from the optics of light, as just mentioned. The positions of maxima and minima on Screen 1 will be determined by the electron's de Broglie wavelength and the spacing of the slits, and may be calculated by the same geometrical methods that were found useful for light. What happened to the particles? They distributed themselves according to an interference pattern, determined by their wave nature.

The first test (one slit closed) demonstrated that the electrons were particles. So did the second test because each electron blackened just one grain on the photographic plate and is responsible only for one point in the pattern whose final intensity is plotted. By the logic of the situation—by our logic—each electron had to pass either through one or through the other slit. Let us, therefore, design an experiment to determine whether a given electron, on its way to contribute to the interference pattern, has passed through one or the other slit. It may be shown, however, with reference to the uncertainty principle that such an experiment cannot be performed.[8] Any test designed to determine the path of the electron as a particle will destroy the interference pattern exactly the way the closing of one of the slits does.

But there is no need to give up. Instead, let us cut down the intensity of the electron beam until there is only a single electron on its way at any time. Leave both slits open and watch the distribution on the screen build up. After a sufficient number of electrons passed the screen, the resulting curve will again be the interference pattern. Next, set up a million or more identical experiments in a million laboratories and record, on a million photographic plates the position of impact of a single electron. After all the films have

Figure 8. Closer examination of intensity distribution of electron interference pattern due to a single slit.

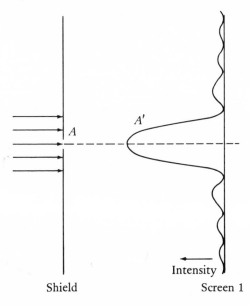

Shield   Screen 1

been superimposed, the combined intensity distribution will again be the interference pattern. Ensemble and time averages will agree.

What the electron responds to is the experimental configuration as a whole. We ask the electron a question: Are you wavelike? It will answer, Yes. We ask it: Are you a particle? The answer is, Yes. One need not even go as far as the double-slit experiment. Figure 8 is an enlarged view of the distribution pattern of electrons that passed through the slit. Examination reveals the existence of side bands, corresponding to a Fresnel diffraction pattern well known from optics and explainable only as an interference phenomenon of waves.[9]

Everything that was said about the electron also applies to the photon. In fact, as mentioned, the double-slit experiment for light was studied a century before it was performed for electrons because, until de Broglie, no one thought of matter as having wave properties.

Matter waves are not to be imagined as those of a physical medium like water; they represent probability distributions; they are probability waves. What may be meant by an object being a probability wave? Anything propagating represents changes in time; in this case the change is the rhythm of probabilities. A probability wave is also a structure in space (witness the distribution of electrons governed by the interference of the structures). Since in the prototemporal world structural and temporal properties are not yet distinct, it must be the case that a "propagating probability wave" is but a fortuitous semiotic sign. It permits the object to be a wave and a particle in its own umwelt, and a wave and particle as well as a wave or a particle, in our umwelt.

We have been talking about free probability waves propagating through space. What happens if one of them, that of an electron, encounters a nucleus? It twists itself around the nucleus in three dimensions and becomes a shell around it, a stationary wave, whose spatial geometry is determined by its wavelength. In the stationary states of probability waves we recognize the quantum-mechanical equivalents of the stationary orbits of the Bohr atom. The electron ceased to be a planet orbiting the nucleus-sun. It is now an object neither spatial nor temporal but both in a way, a probability wave.

Quantum theory handles probability waves as one would handle any other kind of waves: through wave equations. The solutions of the wave equations are the celebrated wave functions of the theory. They make it possible for us to deal mathematically with the primitive variables of the prototemporal world, with phenomena of which we can have no direct experience. By the generalized umwelt principle, the probability waves of quantum theory must therefore be given the same status of reality as we give to length, time, or mass in macroscopic physics.

Einstein, Podolsky, Rosen, and completeness in quantum theory In the tradition of Western scientific thought the quantum-mechanical notions of complementarity, indeterminism, duality, and probability imply incomplete knowledge. They contrast starkly with the well-defined variables of classical and relativistic physics. Yet they are the very hallmarks of quantum theory. In 1935 Einstein and two of his associates gave vent to their intellectual dissatisfaction with the theory in a paper entitled, "Can quantum-mechanical description of physical reality be considered complete?"

In the summary of their paper they set forth their precepts: "In a complete theory there is an element corresponding to each element of reality. Sufficient condition for the reality of a physical quantity is the possibility of predicting it with certainty without disturbing the system."[10] In terms of their own specifications they found quantum mechanics lacking in completeness. We will examine the empirical and theoretical issues involved in their arguments.

Imagine two quantum-mechanical systems, Particle 1 and Particle 2, interact for a limited period of time, and let us assume that the wave function of the combined system is known. Let the two particles separate but assume further that their combined behavior remains describable by the original wave function. This allows us to calculate the state of the total system for any separation of its component systems, even if they are sufficiently far apart to permit each to interact with a separate and independent measuring device.

For instance, after the two particles interact, the momentum of Particle 1 (or at least the momenta of a narrow beam of particles along the flight path of Particle 1) may be measured by Apparatus 1, to any desired accuracy. Or we may measure the position of Particle 1 to any desired accuracy, though not both at the same time.

Because of our knowledge of the combined wave function, if it is the momentum of Particle 1 that we measured, we may calculate the momentum of Particle 2, now at some distance away from Apparatus 1. Likewise, if it is the position of Particle 1 that we decided to measure and have measured very precisely, we may calculate the position of Particle 2. By the Einstein-Podolsky-Rosen definition there is, therefore, a physical reality to both the position and momentum of Particle 2, for they were measured without System 2 having been disturbed. Furthermore, since System 2 could not have known whether it is a momentum or a position measurement that we wish to perform on System 1, Particle 2 must have possessed precise and well-defined position and momentum values all along. Position and momentum—and, likewise, time and energy —must therefore be precisely calculable for elementary objects no less than for macroscopic bodies. But quantum mechanics teaches that this is not the case. The quantum-mechanical description of nature is, therefore, incomplete.

Within a few months after the publication of the Einstein-Podolsky-Rosen paper, Niels Bohr gave a striking reply to the criticism, by summing up concisely the quantum-mechanical view of the universe of elementary particles. The Einstein-Podolsky-Rosen criterion of reality, he wrote, contains ambiguity in the expression, "without in any way disturbing the system."

> Of course, there is in a case like [the example just given] no question of a mechanical disturbance of the system under investigation during the last critical stages of the measuring procedure. But even at this stage there is essentially the question of an influence on the very conditions which define the possible type of predictions regarding the future behavior of the system. Since these conditions constitute an inherent element of the description of any phenomenon to which the term "physical reality" can be properly attached, we see that the argumentation of the mentioned authors does not justify their conclusion that quantum mechanical description is essentially incomplete. On the contrary, this description . . . may be characterized as a rational utilization of all possibilities of unambiguous interpretation of measurements, compatible with the finite and uncontrollable interaction between the object and the measuring instruments in the field of quantum theory.[11]

Consider, for instance, the illustration given. We wish to infer, for the unmeasured System 2, the position and momentum of the particle. Precisely because the system cannot tell which of these variables we will measure, assurance must be had that when the two systems interacted the net momentum and the position of the combined system were knowable, simultaneously, and to any desired accuracy. But such a configuration of parameters and accuracies is prohibited by the uncertainty principle. Quantum mechanics is impressively consistent.

What, specifically, is being introduced into quantum-mechanical measurements when the measuring apparatus is included in the description of the experiment? As a first step toward an answer, we continue by quoting Bohr. He remarked that the

> position of measuring instruments in the account of phenomena . . . appears closely analogous to the well-known necessity of relativity theory of upholding an ordinary description of all measuring processes, including a sharp distinction between space and time coordinates, although the very essence of this theory is the establishment of new physical laws, in the comprehension of which we must remove the customary separation of space and time ideas.[12]

Bohr saw an analogy between the two situations. I believe what we have is a hierarchical continuity, corresponding to the relationship between space and time along the prototemporal and eotemporal integrative levels, respectively. To explain what this means we must take a detour of some length to the notion of space quantization, a concept originating in spectroscopy.

The situation to which the term *space quantization* refers is an extension of quantization to periodic motion, parametrized by the angular momenta and magnetic moments of particles. Quantum mechanics assumes that the angular momentum of an elementary particle is fixed for each species and equals an integral or half-integral multiple of the quantum unit of momentum. However, there is no way to experiment with atomic angular moments except by using the magnetic moments of the particles. It is the direction and motion of the magnetic moment that is being controlled by means of an impressed magnetic field. The direction of the external magnetic field thus serves as a reference for the angular momentum.

Whenever moments of elementary particles are detectable, they are imagined as either being directed along a magnetic field or else being on a conical surface whose axis is directed along the field. The conical surfaces are, however, not arbitrary. They must

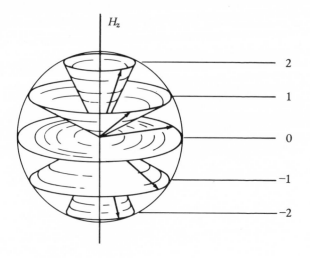

Figure 9. The angular momentum vector is imagined precessing about the direction of the reference field H_z, generating conical surfaces. The length of the generator of each cone, the radius of the sphere, is proportional to the angular momentum of the particle. Quantum theory allows only for certain orientations of the cones. They must be such that the projections of the radius of the imaginary sphere upon the H_z direction assume integer or half-integer values, in units of the quantum of momentum. In the example shown those projections have five values: 2, 1, 0, −1, −2. None of the vectors can point exactly along the H_z axis because that would amount to the simultaneous specification of a spin vector in more than one direction, a condition forbidden by the uncertainty principle. Space quantization designates the existence of restricted spatial directions in this visual representation of the phenomena.

have such central angles as to make the projections of the angular momentum vectors upon the reference direction one-half or whole integer multiples of the quantum unit.

Figure 9 shows a family of coaxial cones useful for the elucidation of the idea of space quantization. The angular momentum vectors are imagined precessing about the magnetic field direction H_z, while remaining within the surface of the cone. The uncertainty principle prohibits the specification of a spin vector along more than one spatial axis simultaneously. Therefore we must think of the vector only as being somewhere within the cone, but not as pointing at a particular instant in a definite direction. Thus, it is the conical surfaces that are quantized. Each cone, including the X-Y plane, represents a different quantum state of the atom, separated from other states by the different energy of its angular momentum. The term, space quantization, designates the existence of preferred surfaces in space in the visual representation of the physical phenomena.

Space quantization is a strictly local phenomenon; it bears no relation to the astronomical world. Also, it does not depend on the strength of the magnetic; it exists in arbitrarily weak fields. But it cannot be assumed to exist in zero field, because zero magnetic field has no direction and how should an atom divine preferred surfaces with respect to a nonexistent reference?

It is evident that what the measuring instrument brings to this quantum-mechanical experiment is a direction; any direction. The prototemporal world cannot define directions in space because in that umwelt space and time are not sufficiently distinct; neither are they continuous.

The name, space quantization, is a misnomer, however. The phenomenon to which it refers does demand a reference direction but it has nothing to do with preferred directions or surfaces, as it was just presented. It is recognized, instead, by the appearance of certain spectral lines separated along a scale of frequencies. Space quantization is not measured by distances or angles but by transition frequencies among energy levels of different angular momenta, and between these levels and the energy levels of other quantum states. While the direction of the external field introduces directedness into the prototemporal world of the atom, the magnitude of the magnetic field, to which the transition frequencies are proportional, introduces a scale factor into the frequency domain. The magnetic field thus helps separate those two aspects of atomic resonance that, in our umwelt, are recognized as spacelike and timelike.

What Einstein, Podolsky, and Rosen missed in quantum physics were Newtonian time, space, and determinism. Niels Bohr's insistence that the conditions of measurements be regarded as "inherent elements of the description of the phenomena to which the term 'physical reality' may be properly attached" is an application of the generalized umwelt principle to the quantum world. For it insists that we think ourselves into the world of the atom and regard its universe as complete and real in itself. The rules for the thinking-into are given by quantum theory.

Our deliberations suggest two points. (1) Quantum theory is complete in the following sense. It renders intelligible a domain of reality which in its modes of connectedness (causation) and in its spatial and temporal features is primitive compared with the causations, space, and time of higher integrative levels. (2) What the instruments introduce into quantum-mechanical tests are spatial directions and time scales.

Delayed-choice experiments   This is a group of experimental suggestions proposed by John A. Wheeler in a series of publications. They are intended to demonstrate that "no elementary phenomenon is a phenomenon until it is a registered (observed) phenomenon" in the world of a human observer.[13] The paradigm and most striking version of the delayed-choice experiment is a cosmic one. It envisages photons starting from a distant quasar. They arrive at the local laboratory along two different paths because of the gravitational lens effect of an intervening galaxy. As shown in the schematic representation of figure 10, the paths will subtend an angle α upon their arrival and should be, therefore, distinguishable.

Since the primary and secondary paths are of different lengths, the secondary signal will lag behind the primary one. To let the later signal catch up with the earlier one, the primary signal is stored in a fiber optics delay loop and storage device. The receiving apparatus consists of two photodetectors and counters, as shown, and a removable half-silvered mirror.

Let the whole system, including the half-silvered mirror be so designed that if two waves arriving along path A and path B are in phase they combine constructively in the direction of Counter B' and destructively in the direction of Counter A'.

When the mirror is out we count photons-as-particles: either A' or B' will register them, as they arrive. With the mirror in we are counting photons-as-waves: when one arrives B' will click, A' will be silent.

According to the reasoning leading up to the experiment, a single photon arriving from the quasar may have come as a particle. In that case it had to come either along the primary path

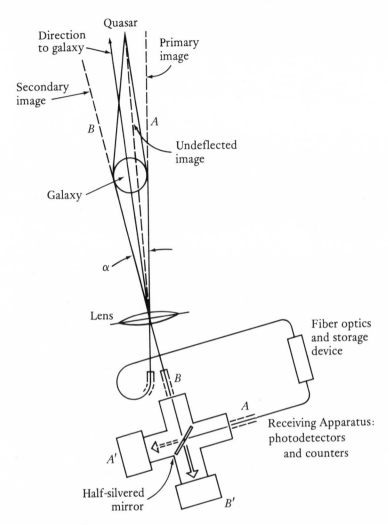

Figure 10. Schematic of the cosmic delayed-choice experiment suggested by John A. Wheeler.

A or the secondary path *B* and will have triggered either Counter *A′* or *B′*. Alternately, the photon may have come as a wave, propagating along both routes. In that case, with the mirror in place, it will be detected in Detector *B′*.

The dimensions of this experiment are vast. Wheeler tells us that we may spend the whole day in deciding what question we will ask the photon, then ask it at night when the telescope is usable. The photons have already been on their way for billions of years. Whether or not all through those years a photon has been wavelike or particlelike, so the argument runs, is decided by our present action freely and unpredictably taken. We are free to determine what is the case that has been. In this experiment

> we are dealing with an elementary act of creation. It reaches into the present from billions of years in the past. It is wrong to think of the past as "already existing" in all detail. The "past" is theory. The past has no existence except as it is re-corded in the present. By deciding what questions our quantum registering equipment shall put in the present we have an undeniable choice in what we have the right to say about the past. . . . Useful as it is under everyday circumstances to say that the world exists "out there," independent of us, that view can no longer be upheld. There is a strange sense in which this is a "participatory universe."[14]

Gravitational lens effects have been observed; images of what are believed to be of the same object have been photographed.[15] The angle subtended by the primary and secondary images is typically of the order of a second of arc, a small angle, but large enough for the images to be separated. The time delay between the two images is, however, quite long, being typically of the order of months or even years. Technology for storing lonely photons for such lengths of time, while also preserving their phase, is not yet on the shelf.

Fortunately, it is not necessary to perform the test because it is but a gigantic double-slit experiment. Its dimensions are impressive but it teaches nothing more than its modest laboratory twin does.

Consider the detecting and counting device with the half-silvered mirror in place. Counter *B′*, the one in which the waves combine constructively, corresponds in its role to the photographic plate at its central, brightest fringe, where the path difference of photons-as-waves coming through slits A and B is zero (fig. 7). Counter A', the one that will keep silent, corresponds in its role to the photographic plate at the first minimum position where the path difference is one-half wavelength.

Give the photon a chance to behave as a wave, it will behave as one. Or remove the mirror and count particles. Each time one arrives, it must be assumed that the alternate path has been blocked. If both click at the same time, you captured two photons from God knows where.

The cosmic delayed-choice experiment demonstrates, once again, the ill-defined separation between processes and structures in the prototemporal world. The eotemporality of the macroscopic devices makes it possible clearly to separate space from time, particles from waves. The biotemporality of the living experimenter introduces the present. The noetic organization of our mind provides the possibility for historic interpretation back a few billion years, or even to the big bang.

But we did not make the photon a particle or a wave retroactively. Such a feat would be comparable to changing the reader into Anthony or Cleopatra, retroactively, by the wave of a half-silvered magic wand. And we most definitely did not introduce reality into the world of the photon. In its own umwelt the photon has been a wave and a particle all along. We just decided whether we ought to sympathize with the Triumvir or with the Queen of Egypt aspect of a hermaphrodite. And as far as the past goes: that is not at all a theory but a fundamental reality of our existence as living organisms. It is true, however, that it can be given no meaning in any of the integrative levels below the biotemporal.

Pair production and annihilation The Wilson cloud chamber is a device used extensively to render visible the paths of elementary particles. The atomic objects themselves remain much below the threshold of resolution of optical instruments. But as they collide with the massive particles of a supersaturated vapor, they leave behind a trail of local condensation as witnesses to their passage, like jet trails record the passage of a jet plane. These paths may be photographed, and from their geometry and a knowledge of the electric and magnetic environment, the physical properties of the particles may be inferred.

It is through the study of cloud-chamber photos of cosmic rays that the existence of positrons was first confirmed experimentally. The properties of these particles had been predicted by Dirac from theoretical arguments. They are the antiparticles of electrons with the same mass, the same spin, and the same amount of electric charge, but they are of the opposite sign. When an electron and a positron meet—if they get closer than about 10^{-13} cm. and remain there for longer than about 10^{-23} sec.—they interact and change into electromagnetic radiation, specifically, into gamma rays. The disappearance of matter in that puff is known as pair annihilation. The opposite process, one that involves the sudden change of a gamma ray into an electron and a positron is also frequently observed. It is known as pair production.

Figure 11 is a schematic depiction of the kind of vapor trail configuration one may observe in a cloud-chamber photograph. The direction of time is represented by the upward pointing arrows. In figure 11a at instant t_1 a gamma ray changes into an electron ($-$) and a positron ($+$). These objects then fly apart. At instant t_2 the positron gets sufficiently close to an electron and remains there for a long enough time to combine with it. The positron and the electron vanish and a gamma ray is produced.

In 1949 R. P. Feynman,[16] following the earlier work of Stückelberg,[17] noted that before instant t_1 and after instant t_2 there is only one world line whereas during the period $t_2 - t_1 = \Delta t$ there are three world lines. He went on to suggest that the history of pair formation and annihilation, as recorded, could be considered in terms of a single electron. We may regard the positron, enduring in forward flowing time to be in fact an electron—the particular electron in question—enduring in backward flowing time. He then showed how this view leads to the simplification of the ways in which certain problems may be stated and for that reason, it has some advantages. He recognized in his suggestion a perspective different from the usual one, which sees the future developing from the past. "Here," he wrote, "we imagine the entire space-time history laid out, and that we just become aware of increasing portions of it successively."

The freedom implied by the Feynman theory is usually interpreted as a time reversal of local nature, a phenomenon, in Whitrow's words, "essentially different from the cosmic time reversal which would be simulated by a film of the world shown in reverse."[18] Whitrow's view may be espoused without also agreeing

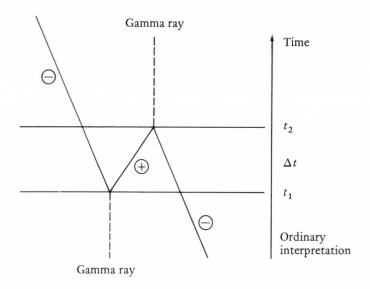

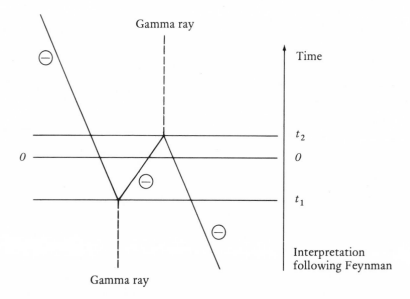

Figure 11. Schematic representation of a cloud-chamber photograph showing pair production and pair annihilation.

to the fatalistic, preordained estimate of the universe and to its necessary timelessness, as implied by the Feynman quote. But I believe that what we are talking about is not a time reversal at all.

In the timelike portion of space-time a single object may certainly be at the same place at three different times. Inspect, however, the line of simultaneity $0{-}0$ in figure 11b. It crosses the path of the same object three times; the same electron is at three places at the same time. A line connecting space-time points in this manner is a spacelike geodesic.

The electron moves in the timelike portion of space-time up until the instant t_1 (on the laboratory clock) and again after instant t_2. During the brief period Δt the electron of the Feynman interpretation has entered the spacelike portion of space-time.

We are reminded of the magnifying glass placed on the light cone in figure 6. The prototemporal world close to the atemporal boundaries of the universe does not distinguish clearly between spacelike and timelike geodesics. If it suits our formalism and fits the boundary conditions, we are free to think of short-lived objects between instants t_1 and t_2 either as an electron moving backward in time or as a positron moving forward in time. A way to describe the situation is to say that for a very short period of time the particle was metastable. In its own umwelt time reversal would make no sense in any case because time there does not have an arrow to be reversed. It does not even have continuity.

Pair production and annihilation are best understood as prototemporal phenomena observed near the atemporal boundaries of the universe.

Preacceleration   In ordinary mechanical problems it is sufficient to specify the position and velocity of a particle at an instant to be able to predict its trajectory. In nonrelativistic electrodynamics, when describing the motion of a charged particle, it is also necessary to specify the initial acceleration. Furthermore, to obtain a physically meaningful solution, it is also necessary to assume that the particle begins accelerating before the force has, in fact, been applied. This phenomenon is known as *preacceleration*. The characteristic periods are of the order of the physical chronon.

Once again we are free to posit a process which, from our point of view, seems to require backward flowing time. We could not do so for large physical objects because although their time is also directionless, their geodesics are definitely timelike.

Zeno's paradox of the flying arrow   Zeno's paradoxes are early examples of thought experiments: they may, but need not, be performed. Everyone knows the answers though not everyone knows the questions. They were formulated twenty-five centuries ago for the purpose of discrediting the belief that multiplicity and change are fundamental attributes of the world, and for promoting the idea that being and permanence are the true bases of reality. We will consider the paradox of the flying arrow.

Think of the flying arrow as occupying at each instant of time a place equal to its own dimensions but no more. For that instant it would be indistinguishable from a nonmoving arrow and must itself be regarded, therefore, as nonmoving. But if an arrow may never be found moving, no matter at what instant it is examined, then it is permanently at rest. The paradox is said to reside in the fact that whereas "The flying arrow flies" is true by our senses, it is false by our logic. If it is the logical path that we trust, we may conclude with Zeno that motion is not a fundamental attribute of the universe.

From Greek antiquity to our own days a vast number of proposals have been made to identify the logical fallacy in the argument or to reconcile its conclusion with our undeniable experience of motion and change. The arguments fall into two major categories.

First, there are those who believe that motion cannot be reduced to rest and hence there is not much use in trying. Second, there are those who oppose this view and believe that either there is, or there ought to be, a way to compound motion from elements of rest and hence it is worth trying.

There is a fundamental tenet upon which members of both of these opposing camps agree. Rest (that is, no change) is a more basic, more elementary constituent of the dynamics of the world than motion (or change) is. It is quite legitimate, therefore, to debate whether or not change and motion are reducible to no change and rest.

But what we have learned about the hierarchy of temporal levels tells us that nature is, in fact, the other way around. Rest and no change are features of the higher organizational levels, motion and change of the lower ones. Light is continually in motion. The atemporal world is intrinsically chaotic, restless; it prohibits any definition of continuity and hence of rest and no change. It is in the prototemporal world that we first find enduring structures though they are still, so to speak, rather flighty. The eotemporal world does have structures whose behavior corresponds closely, but not totally, to our notions of permanence. Only in the mind of man do we finally find examples of absolute permanence, such as those of the Platonic forms. The path of evolutionary development has been from chaos to fragmentation and then to permanence, and not in the opposite direction.

Each epoch in the history of ideas, each major school of thought from the Aristotelians to the Russellians had their favorite ways of making the paradox of the flying arrow go away. This unabated need for the reexamination of the paradox is because the fundamental belief that underlies the premises of the paradox of the flying arrow does not pertain to the real world. We could also endlessly debate whether unicorns come down to the sea at night.

It is possible to formulate questions that are inverted versions of Zeno's paradoxes. These questions, because they correspond to the real world, may be answered. For instance, how can motions be combined to create rest? It is easy. It is done in kinematics all the time: we add two motions of equal magnitudes but opposing directions. How may states of permanence, or close permanence, be constructed from ceaseless change? Cool the universe, permit it to complexify, and produce granite. How may stable scientific laws be formulated, based on measurements of change? By the mental process of induction, deduction, and abstraction.

The fundamental state of the world is motion and change. Rest and no change are not primitive but evolutionary aspects of reality.

4.4 Across the Boundaries of the Prototemporal World

This section traces a number of paths from the prototemporal up to the eotemporal and down to the atemporal universe, noting examples of continuities and discontinuities.

Classical mechanics and Maxwell's electromagnetic theory describe macroscopic phenomena accurately. Quantum mechanics must, therefore, have a boundary along which it blends into the theorems of the macroscopic world. As the classical limits are approached quantum mechanics does not become untrue but irrelevant. This continuity has a name given to it by Niels Bohr; it is called the correspondence principle. To be consistent with the correspondence principle is a formal requirement for all quantum-mechanical laws.

A beautiful example of continuity is the transition from the wave mechanical to the Bohr representation of electronic orbits, and from there on to macroscopic physics. Consider, for instance, the case of the hydrogen atom: nucleus and one electron. Place the atom in an electric field so as to introduce an asymmetry into the electronic orbit. The wave function when plotted (or imagined) in three spatial dimensions looks like an elliptical mountain range with the nucleus at one of the focal points of the ellipse. The height of the mountains at each point is proportional to the probability of finding the electron in that region within an arbitrary but equal unit of time, used to plot the total orbit. The mountains are high at aphelion, low at perihelion. The electron is more likely to be found somewhere away from the nucleus.

Switch to the Bohr representation. It is a single, elliptic planetary orbit about the sun, the nucleus. By Kepler's second law the orbiting electron moves rapidly at perihelion, slowly at aphelion. Obviously, it is more likely to be found away from the sun.

But the proper specification for the orbiting electron is the wave function. The proper representation of the orbiting planet is the Kepler orbit. Nature does not oblige us with an example to which the two descriptions may be applied with equal ease and justification. On this account, at least, there are no mesoforms between the prototemporal and the eotemporal worlds. But yet, the continuity is quite clear.

In some cases the continuity is incomplete, even theoretically, though it is definitely there. Consider, for instance, electron spin, first proposed as an explanation for the existence of unexplained lines in the atomic spectra. One may think of the electron as a particle that rotates about its axis, having both mass and charge, both angular momentum and magnetic moment. Place the electron in a magnetic field not parallel to its spin axis and watch it precess (fig. 9). The absorption or emission of energy at the precessional frequency accounts for the existence of the peculiar spectral lines.

But the spin of the particle is unlike that of a macroscopic body: it is constant and cannot be changed, it is intrinsic to the particle and species-specific to it. It is not possible to find an electron that does not spin or a meson that does. If a macroscopic spinning top is tipped over end to end, its sense of rotation, for a stationary observer, changes. A further tipping of the axis by 180° will bring it back to its original state. A spinning electron must go through two complete tippings before it is back to its original state. There is no exact macroscopic analogue to the spin of the particle even though it is quite correct to say, in first approximation, that it spins. But if spinning is what the earth does, then an electron only behaves as if it had a spin. The only appropriate way to describe it fully is through the quantum-mechanical spin operator. These operators allow us to deal with the prototemporal ancestors of what, in the macroscopic world, we call spin.

Again, nature does not oblige us with systems to which both descriptions could be applied with equal ease and justification.

A large number of similar continuities and discontinuities may be traced between the prototemporal and eotemporal organizational levels of the physical universe.[19]

Let us turn to continuities and discontinuities between the prototemporal and the atemporal levels.

In the 1920s there were many attempts made to combine quantum theory and relativity theory. Theoretical work, designed to overcome certain mathematical difficulties in the equations of matter-wave behavior, led P. A. M. Dirac to postulate the existence of matter consisting of negative energy. By that time atomic physics had revealed the existence of some unusual processes, such as the decay of the neutron: a particle of matter vanishes from the world and another particle of different nature suddenly appears. In 1930, to explain such sudden disappearances and appearances, Dirac invoked his earlier theory of negative energy matter and postulated the existence of an infinite sea of negative energy beneath the lowest energy level of the atomic electron. This infinite sea was assumed to be filled with invisible particles. The positron, whose adventures were discussed earlier, is a citizen of this infinite, invisible sea of negative energy.

Since Dirac's discoveries it is generally held that the vacuum of the physical universe is not empty. Out of it pop and into it disappear numerous structures that complement the temporal structures of the world. Vacuum is filled with messenger particles or virtual field quanta. These are objects that live shorter than the length of the atomic chronon and must, therefore, be regarded as atemporal. Out of the same invisible region come and into it go particles which, at least in theory, need not be short lived. To every particle but photons, gravitons, and the unstable neutral pion there corresponds an antiparticle. They have the same mass as the particles themselves but opposite electric charges, magnetic moments and spins, opposite baryon and lepton numbers. They have the same theoretical lifetimes, the same group of radiation products and, if they could be left alone long enough, they would cascade down to stable particles along the same routes as matter particles do.

Antiparticles are always created together with ordinary particles. Once the two appear they move in opposite directions with high speeds. Matter particles get absorbed and settle down to continue their normal lives. Antiparticles soon get close to matter particles, interact with them, and disappear in the form of radiation, such as photons or Pi mesons, or gamma rays, as in figure 11. Theoretically, antimatter should be stable and could exist even in bulk as long as it does not get close to ordinary atoms. There have been speculations about antimatter stars and galaxies to be found in the immensity of space. There may even be antimatter planets with dialectical materialists made of antimatter, except that there is no evidence whatsoever that antimatter exists in anything but the most fleeting forms.

It is possible to create antiparticles under laboratory conditions and even to make enough antiprotons to be accelerated in a cyclotron. But present understanding of inorganic evolution makes the existence of any significant amount of antimatter in the universe unlikely. To deepen the puzzle, there is no good theory that could explain why matter predominates over antimatter.

The concept itself arose in a schoolbook example of physical discovery: it was theoretically useful, it explained known phenomena, and it made testable predictions. Aesthetically, its great strength is its symmetry. The sea of negative energy particles relieves the physicist from the burden of living with only positive energy particles. The symmetry reminds one of the theoretical basis of astrology: as above, so below. But believing that everything in the universe is necessarily symmetrical is not a defensible stance, especially when empirical data suggests the opposite. The idea of an invisible sea of negative energy particles has been criticized on various bases and, in the words of Devies, "can at best be only a device for visualizing the creation and annihilation mechanism" of elementary particles.[20]

The principle of temporal levels suggests that the relationship between matter and antimatter is hierarchical. Although all formal relations thus far known are those of symmetry, antimatter empirically fits the description of a metastable mesoform between the atemporal and prototemporal integrative levels. This is a conjecture that tells nothing about why antimatter is unstable and matter is stable, but it may help us in the direction of a satisfactory answer.

4.5 The Completeness and Reality of the Prototemporal World

We learned in section 4.2 that the probabilistic nature of the prototemporal world is the obvious evolutionary step between the chaos of the atemporal and the determinism of the eotemporal integrative levels. But it is difficult to come to terms with the intrinsically probabilistic features of quantum theory. The puzzlement of many physicists may be summed up in the following terms.

Speaking the language of wave functions, quantum theory cannot assert anything with certainty, only with various degrees of probability. Probabilistic statements are also often made in daily life regarding future contingencies, but those events eventually either do or do not occur. When it becomes known that the anticipated event did or did not come about, the situation changes from probability to certainty. We may guess that a pointer will point to the numeral 3 or to the numeral 5 but once the reading is made it is either 3 or 5 and not perhaps one or the other. But events in quantum theory remain forever probabilistic. To allow for the empirical differences between the probabilistic nature of the prototemporal world and the certainties of the eotemporal world, quantum theory recognizes pure and mixed quantum states.

Probability waves, when combined by superposition, may yield very complicated waveforms but the new waveform is still a probability wave. We say we are dealing with a pure quantum state. Thus, if the quantum states of all systems making up an ensemble are known, then the individual wave functions may be combined by superposition into a single new one. With the passage of time pure states transform into other pure states according to instructions known as Schrödinger's equations. At all later instants the dynamical state of a system is thus perfectly knowable—in terms of probabilities.

To provide a formalism in quantum theory that can handle the fact that macroscopic readings tell us about certainties, physicists appeal to quantum statistical mechanics. This discipline admits the existence of mixed states, by which is meant a quantum state of wave functions wherein each function has a statistical weight. A different mathematical formalism must then be adopted. A quantum average is first formed. It is known as expectation value. Then the statistical average of the expectation values is constructed. This variable is handled by a statistical rather than quantum operator or, when it is in matrix form, by the so-called density matrix. The nonquantum world may now join the quantum world via statements about statistical averages without disturbing the quantum averages.

We are dealing here with two kinds of probabilities. The probabilistic character of the wave function demonstrates the very nature of elementary objects: they are, as we know, propagating probability waves. An electron cannot be reduced (or rather, elevated) to a thing at a place at an instant. In contrast, the probabilistic character of the quantum average, used in the density matrix, expresses only incomplete knowledge. By performing a measurement we can determine which of the possible quantum averages is present. This latter kind of probability can be reduced to certainty. (Because of the hierarchical levels of causations, one ought to speak about elevating rather than reducing statistical knowledge to certainty!) The use of mixed states, handled by density matrices, acknowledges the fact that a sufficiently populous prototemporal system behaves in an eotemporal way.

It can be shown theoretically that a pure state, in itself, cannot evolve into a mixed state.[21] What, then, may be held responsible for the change of state? In the literature of quantum theory the culprit bears the name, "the measuring process," or words equivalent to it. The principle of temporal levels prefers to recognize that we are dealing with two stable, hierarchically nested integrative levels of nature. The pure quantum-mechanical state represents the prototemporal level, quantum-statistical mechanics represents the prototemporal and the eotemporal levels, together.

Physicists have been disturbed by this implicit but generally unrecognized hierarchy. Thus, the Einstein-Podolsky-Rosen challenge was designed to lead to an eventual elimination of the probabilistic aspects of atomic physics. Einstein's famous remark that God does not play dice shows his conviction that the statistical character of the wave function hides only ignorance. Bohr's insistence that the conditions of measurement be included in the purview of quantum-mechanical description leads us in the opposite direction. It implies that the totality of nature is probabilistic. Both of these opposing stances overlook the hierarchical organization of nature: Einstein's by insisting that the world is (in our terms) eo-temporal, Bohr's, that it is prototemporal.

The problem with Bohr's program is that the measuring device cannot be made the final station along the ever-expanding store of wave functions. We must include the room, the building, our earth with its birds and bees, our galaxy and all that is beyond it. The proposal leads to an infinite regress.

The infinity catastrophe can be prevented only by providing a backstop. One candidate for that position has been the universe because, unlike the atom, it is unique. For the same reason another has been the conscious experience of the physicist.

The relative-state formulation of quantum theory suggested by Hugh Everett III offers a formalism that permits the writing of quantum theory with the universe as its final reference. Everett observes that there is no such thing as the quantum state of a system but only its quantum state relative to the quantum state of the observer (instruments and all). By the Bohr doctrine we must consider the two of them together as the object of quantum-mechanical description. But whereas we allow for the complexification of the system under study by permitting the superposition of wave functions, it has been tacitly assumed that the observing system is always in the same state. The proper way to think of the problem, says Everett, is as

> a representation of terms of superposition, each element of which contains a definite observer state and corresponding system state. Thus with each succeeding observation (or interaction), the observer state "branches" into a number of different states. . . . Each branch represents a different outcome of measurement and the corresponding eigenstate for the object-system state. All branches exist simultaneously in the superposition after any given sequence of observation.[22]

We are asked to imagine a wave function of ever-increasing complexity, always obtained by linear superposition and thus behaving according to Schrödinger's equation. Eventually we arrive at the quantum-mechanical description of the universe. This is our final backstop. The pure quantum-mechanical state that is now the universe is allowed to evolve deterministically, although it is not clear whence the arrow of time might come. At each instant the world itself splits into all possible alternative quantum states which keep on evolving and splitting, independently. How is it then that the history of the universe appears to us as a single series of unique events in a certain sequence?

> [The] "trajectory" of the memory configuration of an observer performing a sequence of measurements is not a linear sequence of memory configurations but a branching tree, with all possible outcomes existing simultaneously in a final superposition with various coefficients in the mathematical model. In any familiar memory device the branching does not continue indefinitely but must stop at a point limited by the capacity of the memory.[23]

The world appears unitary to us because an observer can only be aware of one particular series of instants but not of the immensity of all coexistent alternatives.

In response to critics who could still not understand why ours appears to all of us as so definitely a single continuous world when there are an infinite number of other ones next door, Everett drew an analogy. The earth-centered interpretation of the universe, he wrote, exists simultaneously with the heliocentric description, yet we select the latter as corresponding to fact.[24] But this analogy could hardly be a valid one. The earth-centered description of the universe is frightfully cumbersome and impractical but it remains a possible one. All I need to do in its support is to look out at the stars at night and forget Copernicus. But how can I gather evidence about all the other universes, even in theory, and why is it that all of us happen to agree on this one?

The many-worlds model of quantum theory, as the system is also known, reminds one of the system of cycles and epicycles with which the pre-Copernican astronomers had to contend. That profusion of eighty circles was reduced by Copernicus to forty-eight circles and by Kepler to seven ellipses. It was a paring-down process associated with correcting a few facts and finding the right perspectives. The many-worlds model can be subjected to the same trimming by placing it into the perspectives of the principle of temporal levels.

The need for the right perspective was sensed by Bryce S. DeWitt when he remarked in a review of the Everett-Wheeler interpretation of quantum theory that "the formalism of quantum mechanics makes sense only if there exists a classical (and hence more familiar) realm in which the results of observations can be uniquely recorded and communicated."[25] We can enrich this passing remark with a wealth of detail.

The quantum world remains discontinuous and probabilistic no matter how large it is imagined to be. Continuity does not become reality until after the aggregates of quantum systems have been subsumed in the eotemporal world of physical cosmology. But the pure succession of that world is in no way a classical form of time; neither is it any more familiar experientially than is the prototemporal world. To be able to give the universe a temporal direction it is necessary to acknowledge the functions of the life process. Finally, to give the universe a history—a problem with which both Everett and his critics have been struggling—it is necessary to defer to man.

The evolutionary development sketched makes possible the taking of one actual road out of the infinite number of possible roads. Robert Frost had put it this way:

Two roads diverged in the wood, and I—
I took the one less traveled by,
And that has made all the difference.

The many-worlds interpretation of quantum theory cannot serve as a backstop bringing to a halt the infinite regress implicit in the Bohr program, because it does not acknowledge and it cannot accommodate evolutionary complexification. The quantum world is not the end but only the primitive beginning of evolutionary progress. It is the dynamic change that makes all the difference.

Instead of following Bohr and believing that all physical processes are basically probabilistic we may decide to begin walking in the other direction. A backstop to the infinite regress of quantum theory will be recognized when we arrive at considering the conscious experience of the physicist. In this view, it is consciousness that changes the quantum statistical probabilities to the certainties of the measured values. By a popular shorthand notation, consciousness is held responsible for the collapse of the wave packet.

This approach has been championed by E. P. Wigner. He believes that it is impossible to give a description of atomic phenomena without reference to consciousness. Therefore, he postulates that "there are two kinds of reality of existence: the existence of my consciousness and the reality and existence of everything else."[26] Based on this postulate he builds an understanding of the position of scientific truth in the scheme of things.

Dualistic theories of this kind, under the heading *mind-body problem* have been of interest to thinkers since Plato. He sharply distinguished between mind and body and held that the mind could exist both before and after its residence in the body. Consciousness in the dualistic interpretation of quantum theory generally stands for the concept of the mind. I intend to maintain that the dichotomy, consciousness versus everything else, is too hasty a conclusion to draw from the teachings of quantum theory.

Each time a micrometeorite impacts on the moon all kinds of wave packets collapse. In detail: there is a change, a simultaneity of chance, which cannot be described by the superposition of earlier wave functions but only by a density matrix. In it, the statistical average of the quantum average must be replaced by a finite, well-defined quantity. As we have reasoned all along, simultaneities of chance constitute the events of the physical world.

The collapse of the wave function does not in itself signify a present, however; the umwelt of the moon remains eotemporal. If the collapse of a wave function corresponds to the arrival of a photon upon the retina of a leopard, and if it leads to inner biochemical changes or overt behavior necessary to maintain life then, and only then, can we speak about a meaningful present. But still, there is no need to appeal to consciousness or imagine that life has introduced reality into the physical world. It only introduced biotemporality.

If the quantum statistical uncertainty is elevated to certainty in the mind of a physicist then, and only then, and only to the extent that this certainty becomes part of his self-awareness, will it be necessary to defer to conscious experience.

The road taken by Robert Frost expressed his human freedom. The road taken by the mouse is determined by the cheese. The road taken by the moon is one of a series of chance coincidence, and it goes both ways. Particles take no roads because in their umwelt nothing can correspond to the metaphor.

Quantum theory does not teach the duality of the world or imply a bifurcation of reality. When seen in the perspective of all modern science, it finds its place as a theory of elementary objects.

It has been said by one wiser than most of us, give God what is God's and Caesar what is Caesar's. The principle of temporal levels recommends that the prototemporal integrative level, but not more nor less, be given to quantum theory, to be governed according to its probabilistic laws.

Thermodynamics

Two Opposing Arrows of Time

The purpose of this chapter is to demonstrate that all thermo-dynamic systems smaller than the cosmos are eotemporal.

5.1 The Thermodynamic Arrow

In *The Nature of the Physical World* Eddington proposed to identify the physical basis of the human experience of time. He coined the phrase *time's arrow* to designate the "one-way property of time which has no analogue in space." This one-wayness is singularly significant because "it is vividly recognized by consciousness [and] is equally insisted on by our reasoning faculty, which tells us that a reversal of the arrow would render the external world nonsensical." The arrow "makes no appearance in physical science except in the study of organization of a number of individuals."[1] By individuals is meant elementary particles believed to exist in a timeless world. By number of individuals is meant the aggregate of elementary particles believed to exist in a world of future, past, and present.

We learned in section 3.6 that the laws of classical kinematics and of electromagnetic propagation are indifferent to a direction of time. "There is only one law of Nature—the second law of thermo-dynamics—which recognizes a distinction between past and future more profound than the difference between plus and minus." The variable of the second law of thermodynamics, the measure of the collective organization of individuals which helps us identify the arrow of time, is a quantity known as *entropy*. "Nothing in the statistics of the assemblage can distinguish a direction of time when entropy fails to distinguish it." But fortunately, it does.

Without any mystical appeal to consciousness it is possible to find a direction on the four-dimensional map [of the world] by a study of organization. Let us draw an arrow arbitrarily. If as we follow the arrow we find more and more of the random element in the state of the world, then the arrow is pointing towards the future; if the random element decreases, the arrow points towards the past. That is the only distinction known to physics. This follows at once if our fundamental contention is admitted that the introduction of randomness is the only thing which cannot be undone.[2]

The Eddingtonian vision is more than a clear statement of a theory which, on its surface, appears to be unassailable. It shows the deep conviction that one physical law or another must eventually lead to time's arrow. "There is only one law of Nature" means one law of physics, negating by implication any other law of nature external to the concerns of physical science. Since time's experiential arrow is undeniable, its sources must be found in that law.

Eddington's program—for it is a program rather than a detailed theory—has been submitted to keen criticism by many scholars and scientists and corrections to it have been applied here and there. But the underlying belief that the direction of time must have its roots in physical process has been retained dogmatically across the literature of time in physics. And the most broadly accepted theory is still Eddington's thermodynamic arrow.

5.2 The Concept of Entropy and Its Many Forms

The rich notion of entropy first arose in connection with attempts to establish a theory of heat based on empirical laws of heat transfer. The origins of the need for some such concept may be traced to the belief by the early nineteenth-century founders of thermodynamics that nature displayed certain large-scale trends. For example, Sadi Carnot, in his celebrated work on the motive power of heat (1824) observed that although energy may be conserved during physical change, some of the energy will be rendered unavailable for more work. This observation became the basis of the second law of thermodynamics. The law itself took shape only gradually through the work of Rudolf Clausius, William Thomson (later Lord Kelvin), and others.

It has been known that heat does not flow from a colder to a warmer body. Therefore, it is not possible to use the heat content of a body to do work if the body is colder than its environment. To give a formal expression to this fact, in 1854 Clausius introduced a new concept which he called entropy. The word was derived from the Greek τροπή, meaning transformation. For Clausius it meant *Verwandlungsinhalt,* the "transformable content" of a system's heat, contrasted with the energy of the system, from the Greek ἐνέργεια, which he assumed to signify *Werkinhalt,* or "work content." Entropy as a variable was then so constructed as to be a measure of the transformable energy content of bodies. It was defined to be numerically equal to the heat gained or lost by a body, divided by the absolute temperature of the body. Simple calculation will show that any flow of heat from a hotter to a colder body—an ice cube melting in a glass of hot tea—results in an entropy gain of the combined system. This nondecrease of entropy, corresponding to the experience of the one-way flow of heat, is the basic teaching of the second law of thermodynamics.

In a series of papers in the 1870s Ludwig Boltzmann reinterpreted the second law as a combination of laws of classical mechanics and probability. He demonstrated its essentially statistical nature. Entropy now appeared as a measure of the probability of a system being in a specific state, with respect to a better ordered state of the same system, used as a reference. Boltzmann's H-theorem, as his proof is known, allowed the rephrasing of the second law. In the course of time, according to the law in its new form, all physical systems isolated from their environments evolve toward more probable states.[3]

A further extension of the notion of entropy and of the second law took place with the development of information theory. This theory employs numerical measures of information gained, or lost, as the contents of a message are learned or forgotten. The unit of information is the *bit,* for binary digit. A bit is the specification of choice between two alternatives: left / right, yes / no; Chrysler Building / Empire State Building; male / female; negative / positive. The information content of an event is not intrinsic in the event. It is determined, instead, with respect to a possible alternative event: go / stay; arrived / did not arrive. Likewise, the information content of a structure is not intrinsic in the structure. It is determined with respect to other, possible, alternate structures.

In Boltzmann's H-theorem entropy measures the degree of disorganization with respect to a well-organized reference. In information theory, information is a measure of an event or structure in terms of what it could have been but is not.

Information theory recognizes an isomorphism between the flow of heat and the flow of information. This isomorphism is so complete that for appropriate choice of constants the relation $\Delta I = -\Delta S$ precisely holds. (I is information, S is entropy in appropriate units.) Entropy with a negative sign, following L. Brillouin, is called negentropy. The second law of thermodynamics may now be read as follows: all physical systems isolated from their environment evolve toward states of decreasing information content.

Entropy and entropylike variables have, by now, been constructed for a very large number of different processes and structures in physics, in biology, and in the behavioral sciences. Concepts of entropy and information have been reformulated in many different ways so as to make them suitable for the description of broadly different phenomena. Furthermore, since entropy cannot be directly tested or even felt—unlike mass, temperature, or force—it was also necessary to identify new theoretical and empirical relationships through which they may be measured. Gases in containers still illustrate the phenomenon of entropy and information flow on the pages of textbooks and understandably so. But the simplicity of the classic examples had long been left behind.

The dying Hotspur in Shakespeare's *King Henry the Fourth* tells us that time "takes survey of all the world." Indeed it does, and not only conveniently simple portions thereof. If Eddington's program is taken seriously, we must include in the purview of the second law *all* physical, biological, and social processes, in a statistical way and respecting, of course, the intent of the proposal. We are not obliged to carry out all possible calculations or get lost in irrelevant minutiæ. But we are obliged to explore the suggestion systematically so as to convince ourselves of the soundness and universality of the claim.

A good way to do this is to focus on the fact that both entropy and information are referential quantities like potential energy. We cannot talk about the intrinsic entropy or information content of a system in itself any more than about the noise of one hand clapping. Therefore we will ask, How is this need for referencing manifest in the formulation of entropylike variables along the different integrative levels of nature? Once we have the answers, we can begin integrating them into the overarching claim of the second law of thermodynamics and time.

In the world of the primordial atemporal chaos, no meaning can be given to comparisons between states. There can exist no definable states and no degrees of ordering. Since an atemporal universe is intrinsically uncorrelated, its entropy or information content is indeterminate. This fact entitles us to assign to the atemporal cosmos whatever entropy or information value seems appropriate so that it may serve as the correct boundary condition to the temporal universe.

Everett's quantum-mechanical cosmos may be the correct model for the prototemporal universe just out of chaos. We imagine it to be in a pure quantum state. It is not possible to take a historical path in it because time, space, and causation are fragmented; all roads in it coexist. What the coming about of space, time, causation, and matter guarantee is the possibility of ordering.

Baryon is a collective name for nucleons such as protons and neutrons. These are the first structured forms of energy believed to have been created by the big bang of the atemporal chaos. Physicists have been giving numerical values to the entropy of baryons. They reason that when cooked in hot stars each of them (the protons, not the physicists) creates about 10^7 photons and increases the disorder of the universe by an amount of entropy proportional to that number. If the currently accepted value of 10^{80} for the baryon content of the universe is close to fact, 10^{87} may be a number characterizing the entropy content of the prototemporal universe.

This or some other figure may then represent the "transformable energy content" of the universe with some nontransformable energy possibly hiding somewhere. But there were no baryons in the atemporal world, for in that universe all objects had to travel at the speed of light. Baryons had to come about in an entropy-decreasing, information-increasing phase of creation. The meter rod of section 3.5, which was made of a radioactive substance that was decaying today, had to have its molecules manufactured from radiative energy some time in the past. We thus have a hidden reference to the entropy value of the universe: the number 10^{80} assigned to the atemporal cosmos.

Consider next that the entropy of a pure quantum mechanical state is conventionally taken to be zero because it is a state as highly organized for its kind as it can possibly be. The same must hold for the universe if it is in a pure quantum mechanical state, obtained by a linear superposition of the wavefunctions of all of its components. When a macroscopic device is used to perform a measurement on a pure state and thus produce a mixed state, the entropy of the system is assumed to increase. This is so because a degree of disordering now replaced the original level of ordering. The act of measurement amounts to a hierarchical referencing: the instrument introduced the eotemporal forms of distance, time, and causation. The same may be said about the appearance of ponderable mass as the universe began to cool.

We now have a curious thermodynamic scenario for Creation. "In the beginning" there is the chaotic atemporal world with its indeterminate entropy content. Somewhere "along the line" the ordering occurs: baryons and other permanent kinds of structures are created. At some instant the system was as highly organized as it could possibly be with a zero entropy content assigned to that state. At some instant the entropy content of the universe shot up to 10^{87} on its way to higher values, obliging us to assign the entropy of 10^{80} to the original state. At a leisurely pace, in 10^{11} seconds the universe became matter dominated, beginning its development toward the laboratory instruments of twentieth-century physics.

The creation of the prototemporal universe is said to have begun at the end of the so-called Planck time, which is 10^{-43} seconds. By the time 10^{-6} seconds have elapsed the proton-antiproton pairs, created earlier, began their mutual annihilation.

Instead of trying to place these hypothetical events—and some more—into a chronological order, one that would make no sense in any case, it would be more honest as well as more accurate to look at them as happening simultaneously. The suggestion comes to mind: the instant of Creation should be regarded as the simultaneous coming about of orderability and disorderability. It is not order or disorder that cosmic creation produced out of chaos but conflict, and the possibility of evolutionary development of complexifying modes of conflict.

I will maintain that ordering and disordering, defining each other, not only emerged simultaneously from the atemporal chaos but have remained two aspects of the cosmic process. Together they are responsible for the evolutionary character of the universe.

What do entropy considerations look like in the macroscopic world? It is there that the whole idea originated, as we have seen. The reader may treat himself to examples of the practical use of the entropy concepts by examining the intricate temperature vs. entropy tables and graphs of a good mechanical engineering handbook. These are employed in the design of all equipment using steam.

For the purposes of the tables the entropy of a system is arbitrarily taken as zero at $32°$ F, for a prescribed value of pressure. Cosmological considerations are not necessary for the building of steam engines. Neither are they important for the laboratory where a perfect crystal at zero degrees Kelvin is assigned zero entropy because it is assumed to be perfectly ordered. Both choices are practical and rational; both systems can go only downhill if isolated from their environments. But in the larger scheme of things both systems had to achieve their particular state of organization either by having deteriorated from even more highly ordered states, if possible, or else by having increased their degrees of ordering from a prior, lesser level of organization.

Stepping from the eotemporal up to the biotemporal integrative level the question is, To what kind of sheep should we attach zero entropy? To the perfect sheep, of course, but it is difficult to agree on specifications. Based on what we know about organic evolution there can be no such thing as a final, perfect form of an organism.

To get around the difficulty posed by the absence of a paradigmatic state of ordering, compared to which a living organism could be said to be disorganized, we may take advantage of the relationship between negentropy and information. We may then use as our reference an arbitrary organizational level much below the actual level of organization of a living system. Information theorists in biology speak about measures of specificity rather than information or negentropy. To find out the measure of specificity of a sheep one picks a base line, perhaps on the cellular level, and begins counting the number of binary decisions that nature had to make to arrive at the sheep. The numeral so obtained is the information content of the sheep in bits, with respect to the arbitrary base line. Ideally, the reference should be on the level of elementary particles but that would make the computation too awkward.[4]

Because time takes survey of *all* the world, we cannot stop our considerations with the living functions of matter but must continue to the integrative levels above the biological. The noetic and social organization of man surpasses in its capacity for ordering, and hence of degrees of disordering, anything of which mindless life or inorganic matter are capable. This is so because we can create mental images of nonexistent structures and processes and through individual and collective behavior construct them from available forms of inanimate and living matter. We thus have the arts and letters, technology and the sciences. How are we to measure the information content of mental work?[5] It is not enough to be able to give the information content of a book in terms of the alternatives that its printed letters could have displayed. The computation should include the development of human speech and writing and of the human mind itself. It is not an easy task.

The necessary change in variable from entropy to information, as we cross from the physical integrative levels to the biological one, is explained by Davies as follows: "The law of entropy increase refers to closed systems, the law of information increase to open systems."[6] (An open thermodynamical system is one that can maintain its identity through time only if it can exchange energy and/or entropy with its environment [see sec. 5.5].) The change in variable is a very significant step; it is not simply a matter of convenience such as changing from ergs to joules when measuring work or from dynes to newtons when measuring force. Using information rather than entropy as the variable is made necessary by the fact that for living systems an ideal reference model cannot exist, whereas for physical systems it is possible to construct such models.

In its turn, the difference has its roots in the different temporalities of the physical and biological integrative levels. Whether the causation is probabilistic or deterministic (whether we deal with a prototemporal or an eotemporal system) the final state of a closed physical system may be envisaged and even calculated, at least in principle. A reference state may thus be defined. The characteristic connection of the biotemporal integrative level is goal-directedness determined by the needs of the organism in its physiological present. Goals are contingencies and not necessities, they may or may not ever be reached. The final state of living organisms as individuals and species, or of the kingdoms of animals and plants, cannot be envisaged, let alone calculated. A final reference state is meaningless.

With these remarks upon the many forms of the concept of entropy, we turn to note some of the received teachings of thermodynamics.

5.3 The Three Laws of Thermodynamics

Each of the three laws of thermodynamics has something important to contribute to an understanding of the genesis and evolution of time.

The first law is a statement of the principle of conservation of energy. It asserts that the net flow of energy across the boundary of a system is equal to the change in the energy of the system. In different words, the internal energy of a system cannot be changed by internal processes be they mechanical, electrical, chemical, biological, or, presumably, psychological. Energy change may come about only by receiving energy from the outside or losing it to the environment.

The formulation of this principle assumes the existence of a larger system, of which the system under consideration is a part. Thus, if we confine our attention to a system closed off from a larger environment or else to the universe with a fixed amount of energy, it is not possible to say anything about an arrow of time using energy as a variable. But such a system need not be changeless. The claim is only that by measuring global energy content, one cannot arrive at directed change. A closed system from the point of view of its energy is eotemporal.

Let us approach the second law by considering its early meaning for steam engines. There is a difference between (1) the nature of the internal energy of the steam engine (the random, heated rush of the molecules) and (2) the nature of mechanical energy (the macroscopic, ordered motion of the machine, such as the rotation of a wheel).

Imagine a cylinder filled with molecules madly running in all directions. The usefulness of the machine is this: the kinetic energy of the molecules gained from burning coal can push the piston now this way, now that way, driving the wheel through a linkage. Because at any one time only a portion of the molecules is likely to rush toward the piston, one would not expect all the kinetic energies of the particles to change into macroscopic kinetic energy. But there is yet a more subtle limitation to the efficiency of the transformation. As the molecules that rush against the piston collide with it, their parallel flight becomes random. This loss of ordering, this increase of entropy is part of the price paid for moving the piston.

We have employed here two forms of descriptions: of the molecules and of the piston. Both registered changes while the energy of the system remained constant. These two kinds of descriptions are the eighteenth-century analogues of what today are called fine-grained and coarse-grained specifications of physical systems. Going from the first to the second amounts to crossing over from the prototemporal to the eotemporal integrative level. Although the increase of disorganization takes place in the prototemporal world, it is measured by eotemporal means. The work delivered by the steam is compared with work believed to have been theoretically available, calculated from prior, macroscopic measurements. The inefficiency is accounted for by the entropy increase.

The third law of thermodynamics asserts that absolute zero temperature may be approached without limits but never actually attained. The best one can do by way of cooling even an ideal gas, is to repeat a two-step process: isothermal compression followed by adiabatic expansion. One cycle can cut the temperature of the system into half. To reach absolute zero an infinite number of operations would be needed.

Absolute zero temperature is not characterized by a complete absence of motion or energy from the system. At $0°$ K the lattice of a solid is not in static equilibrium but performs what is known as zero-point vibrations. The particles vibrate about the position they would occupy if they were at rest. The situation is a consequence of the uncertainty principle. If no meaning can be attached to the notions of a definite position at a definite instant, then we are not entitled to imagine an object as being at rest at an instant. The elementary object is spread out, having remained a propagating probability wave. Thus we must revise the earlier description of a particle vibrating about a point. What one has is minute probability clouds arranged in space about imaginary centers. Within these clouds the world remains intrinsically probabilistic. Absolute zero as a macroscopic state joins the atemporal border regions of the universe. It represents yet other conditions that objects of the temporal world can only approach but not actually achieve.

Although, as we have learned, there is no single method for the measurement of entropy or information across all the integrative levels of nature, the principle of the second of the three laws of thermodynamics carries great conviction. It speaks about decay as the truly universal trend in nature. In senility, which is the decay of the noetic faculties of the mind, mental functions relapse into biological ones; the patient vegetates. In death, living organisms return to inorganic matter. In the eotemporal world mountains and even stars wear away. In the prototemporal universe complex molecules break into their constituent atoms and even matter itself, unless held together by internal forces, relapses into atemporal radiation.

Before this impressive universality of the principle may be put to use as a guide to the physical sources of the arrow of time, two conditions must be fulfilled. Both of them have already been mentioned. First, the system must be part of a larger system with which it does not exchange matter, energy, or entropy. Second, the system must have an initial state sufficiently organized so that a loss of order is possible.

These two conditions lead to a number of questions. For instance, how are we to think of the universe from the point of view of the second law of thermodynamics? That object is not part of a larger system, by definition. Or, consider the fact that most processes do not begin with a high degree of ordering and then just run down. In the hot interior of stars heavy elements are continuously being formed from light elements by nucleosynthesis. Ordering is created but so is disordering through the emission of energy in form of photons and neutrinos. The expansion of the universe itself creates ordering, such as the structural grouping of galaxies. Meanwhile, everything is subject to decay. A horse is in continuous decay as are all readers of this book, while at the same time both horses and people are in continuous self-organization. How are we to think of the entropy or information content of a complex dynamic system in which the decay and growth processes are so intimately intertwined that they could not be separated even in an idealized experiment?

5.4 Reversibility and Irreversibility

Implicit throughout the literature of time in physics are two universal, metaphysical beliefs concerning irreversibility and time. According to the first one, there must exist at least one physical process that is truly irreversible. According to the second one, once such a process has been identified it will either itself be found responsible for the one-wayness of our experience of time or else it will lead us to the physical process which really is. These two pleasing articles of faith appear to be quite clear, but they are not.

In his inaugural dissertation, Max Planck remarked that it was not sufficient to define an irreversible process as one that cannot run backward because—although a process may not normally run backward—the initial conditions could, nevertheless, be somehow restored. What was necessary, he said, was for nature to have a preference for the final state.[7] We will change the focus of this definition from states to processes and make the distinction sharper.

An irreversible process is one for which nature has such a preference that the reverse process becomes meaningless.

To illustrate what this means I will sketch four thought experiments and ask, for each, whether its time-inverted version is or is not possible in the scheme of nature. The discussion will not be concerned with what happens to the entropy or information content of the laboratory, the solar system or the galaxy as the experiments are performed.

—Prepare a bell jar. Pump out the air through a hole in its floor. Connect the hole to a perfume bottle. Open stopper, photograph container filling with perfume molecules. Stop film when constant pressure is reached. Project film in reverse and ask vacuum-system engineers to describe and explain what they have seen. Conclude that there is nothing impossible about the time-reverse process of diffusion. Therefore, diffusion is not irreversible.

—Prepare a copper tube. Pass electricity through it and observe the warming of the environment. Photograph the increasing agitation of the molecules surrounding the tube. Project film in reverse and ask refrigerator engineers to describe what they have seen and explain how it was done. Conclude that there is nothing impossible about the time reverse of a heating process.

—Take a new deck of cards and shuffle it. At unit intervals of time lay out the cards in the sequence they happened to be. Make still photographs of the distribution until expert shuffler says cards have been thoroughly mixed. Copy the stills on a moving picture in the sequence they were made, leaving each distribution on the screen for a few seconds. Project film in reverse. Ask card manufacturers to describe how they assemble their decks of cards from what originally are different piles of cards printed, delivered, and stored separately. Conclude that it is not at all impossible to produce an ordered set of playing cards from a disordered one. Card mixing is not an irreversible process.

—Prepare a time-lapse film of a tulip growing from its bulb to full flower. Project film in reverse. Ask anyone but science-fiction writers to explain in theory or demonstrate with experiment the ungrowth of the tulip.

What could be meant by the ungrowth of a tulip, in any case? It would be a stage-by-stage divesting of the organism of its mature structures and functions. It would have to be controlled by a program using the kind of internal coordination we called simultaneities of need and it would have to lead, eventually, to the bulb state.

The celebrated growing-ungrowing tulip will have to have in its genes some new instructions. These must provide for a phase of ungrowth, a series of states subject to the same environmental challenges as the stages of growth are. The genetic endowment of the new plant will have to be more complicated and more extensive than the genetic endowment of the tulip that does not ungrow. The ungrowth will have become part of the growth process.

There is no need to experiment with growing-ungrowing tulips, however. Nature has already done the experiment with much more finesse. The individual ages and dies but produces seeds. As far as temporal umwelts go seeds represent a regression; they correspond to ungrowth. Once upon a time Plato, Aristotle, and even Sir Isaac Newton were no more than giant molecules. But seed production, though a regression morphologically, is part of the growth process of the species.

Applying Planck's criteria to these four examples, it is only the life process that proved to be irreversible. Nature has such a preference for the life process that it subsumes its reverse process in the forward process. A time-reverse process of life is meaningless. Living things do, of course, die. But a dead body decomposes; it does not ungrow. Irreversibility may thus be seen as being closely tied to the necessity of internal coordination, to the maintenance of the identity and autonomy of a dynamic system, to the physiological present. At least as far as these arguments go, no inorganic process can be irreversible.

This fact does not make irreversibility less of a natural phenomenon. It does not make it mystical, distasteful, unreal, or impossible to subject to scientific inquiry. It only removes irreversibility from the immediate concerns of physics.

As promised, we did not discuss how the entropy or information content of the universe might have changed while perfume molecules came and went, copper tubes heated and cooled, card decks got shuffled and assembled, and tulips grew and reproduced. To be able to consider these problems it will first be necessary to expand the spatial and temporal domains of our interest.

5.5 Closed and Open Cases

Idealized maps of real structures are known in physics as models. They have been found immensely useful. A model is usually retained faithfully as long as predictions based on it remain consistent with experimental findings—more or less. A *thermodynamically closed system* required to make the first and second laws of thermodynamics valid is such a model.

The theoretical boundaries of a closed system may be a mathematical surface. An actual and almost ideally closed system may be a box one hundred astronomical units across, placed into empty intergalactic space and slaved to the motion of a free watermelon at its center. Except for the watermelon the box is to contain only vacuum. The walls of the box must not boil off in high vacuum but they must shield the inside from electromagnetic fields. If this project is too expensive, we might imagine a more modest surface enclosing only the earth. Or, hope for a well-insulated box on the laboratory table.

It is to boxes such as these that the entropy increasing, information decreasing provisions of the second law apply.

A *thermodynamically open system* is one surrounded by an imaginary or real surface across which matter, energy, and information may freely flow. The need for formulating the idea of open systems arose from the peculiarities of self-organizing structures. If self-organizing systems are prevented from exchanging energy, matter, or entropy with their environment, sooner or later they will cease to be able to maintain their structural and functional identities and thus their integrities.

The growth of vertices in certain viscous fluids are self-organizing and, therefore, must be regarded as thermodynamically open. The same holds for the moving striations in the positive column of gas discharge and for the lining up of atoms in lasing action, just before the emission pulse. Recently there has been an increasing interest in certain turbulent flows, also thermodynamically open. Such processes are sometimes called dissipative and may be analyzed through chemical kinetics.[8] They seem to be some of nature's attempts to create self-organization. But, for the creation of stable dissipative systems, as we shall learn in section 7.1, evolution has taken a different route of development. The end product of that development is the life process.

Biological systems absorb matter and energy from their environment, then return them in biologically less useful forms. While organisms are alive they create increasing order inside their envelopes, at the expense of the world outside. The entropy of the living system changes by internal processes and also by interaction with the environment. The entropy of the living system may decrease with time because the rate of outflow of disorganization is larger than the rate at which the disorganization is produced.[9]

To the living organism itself, taken as a single structure within a permeable membrane, the entropy increasing provisions of the second law do not apply. On the contrary, as Needham remarked, writing about life and organic evolution: "The law of evolution is a kind of converse of the second law of thermodynamics, equally irreversible but contrary in tendency. We are reminded of the two components of Empedocles' world, φιλία, friendship, union, attraction; and γετκοs, strife, dispersion, repulsion."[10]

If we take our reference in the living system then our sense of time parallels the decreasing entropy, increasing information trend in nature. Let this be called the growth arrow. If we take our reference in the large closed system that includes the living organism as an open system, then our sense of time parallels the entropy increasing, information decreasing, or decay arrow. These arrows are defined, one with respect to the other. We may speak about the direction of either only because there are two opposing trends. Upward makes no sense unless there is a downward.

The belief in our sense of time as necessarily paralleling the decay arrow is a very deeply seated one. How could such a clearly erroneous belief have become an unquestioned dogma? There are good historical reasons for it, just as there were excellent reasons for believing at one time in a geocentric universe.

(1) Our acquaintance with closed systems preceded by a century our familiarity with open systems. Understanding in physics is professed to be ahistorical but, in fact, it is based on conservative precedents and conventions (see sec. 8.1).

(2) We are immersed in an immense inorganic world. The biomass on this earth, measured in tons, is totally negligible in comparison with the mass of the universe. The behavior of living matter appears totally inconsequential for any great issue, such as the question of time, to a kind of thinking more impressed by the facts of physics than those of life, mind, or society.

(3) Western civilization is heavily mechanistic. Our clocks (see sec. 1.1) are exclusively physical. Were our timekeepers organic, were they to keep time by

> Hot lavendar, mints, savory, marjoram;
> The marigold that goes to bed with the sun,
> And with him rises weeping . . .

rather than by clocks ticking and tocking and winking, then our sciences might have developed along different lines. We would have come to believe long ago in time's arrow as pointing in the direction in which the information content of living organisms increases. Learned papers would have noted that the inanimate world, with its lifeless stars, is running in the opposite direction. Someone might even have remarked that it made no sense to insist on time as paralleling the direction of the growth arrow unless simultaneous reference is made to the decay arrow.

The association of the direction of time from past to present to future with either the growth or the decay arrow is arbitrary. It depends on whether one prefers to side with the unpredictability of the living or the certainty of the dead.

All self-organizing open systems, if they are to retain their identity, must be parts of a larger system. By the relative dimensions of living organisms, the larger system is most likely to be a closed, inorganic one. It may be shown that any decrease in the entropy of a living system is overcompensated for by an entropy increase of the closed one, containing it. This fact gives added weight to the decay-arrow argument of time but only in tons, not in logical status. For it does not alter the arbitrary nature of choice.

Imagine now that the closed box gets enlarged from its solar-system size to the size of our galaxy. A physicist, wisely remaining outside the box, may well insist that the final state of the box will be one of thermodynamic equilibrium and with its arrival will arrive the end of time. Meanwhile, however, he is free to select the direction of time to parallel the decay arrow of the box, or the growth arrow of his own biological system, or that of his horse, which he brought along for transportation.

Feeling befuddled by the issue of entropy and time, he is likely to decide to enlarge the box further until it becomes coextensive with the universe. His decision will lead to curious consequences. The box of formerly stable dimensions will change into a container whose size varies, as judged by his sense of time. Also, because there is no place left outside the box, physicist and horse have to hurdle the enclosure and jump inside.

Once safely at his desk, the physicist will be well advised to put thermodynamics aside for the moment and study, instead, the general theory of relativity.

General Relativity Theory

The Natural History of the Eotemporal World

General relativity theory constructed a cosmology to which life and man appear to be irrelevant. By so doing it has demonstrated that the time of the physical universe is one of pure succession.

6.1 The Two Theories of Relativity

It is customary to distinguish between the special and the general theories of relativity by claiming that, whereas the special theory deals with the physics of frames in uniform translation, the general theory handles the physics of accelerated frames. But the formalism of the theory, as we learned in section 3.5, is quite adequate for the transformation of accelerations according to the principle of relativity.

What the special theory of relativity cannot do is deal with the behavior and effects of matter. Whereas the earlier theory concerns the physics of empty space-time, the general theory deals with the physics of space-time containing gravitating masses. Because the universe is held together by gravitation, the general theory is one of cosmology. It does not widen the earlier theory but generalizes, instead, a particular hypothesis concerning the nature of space-time. Through the use of certain notions of geometry it constructs a model of the universe whose features resemble one derivable from classical mechanics much more closely than anything in the special theory of relativity resembles the corresponding teachings of Newtonian physics.[1]

Among the basic differences between the Newtonian and general relativistic models we find the two kinds of temporalities that they deem to be appropriate for the study of the cosmos. Whereas the Newtonian universe is nootemporal, the general relativistic world is eotemporal.

6.2 The Astral Geometry of Gauss

Need for the quantitative organization of spatial forms has surely arisen in all early civilizations. The origins of geometry, according to Herodotus, reach back to the ancient Egyptians who invented it for the purposes of measuring land. Democritus wrote about Egyptians who were rope-stretchers or, in Greek, *harpedonaptai*. This was an activity employed for land survey, architectural work, and astronomical determinations, having been in use as early as 2300 B.C. A stretched rope was then, as it is now, a practical way of determining the shortest distance between two points. By definition, such a distance is a straight line.

Our concerns with relationships among shapes probably has its origins in our search for order among kinesthetic experiences. Whatever its origins, from Euclid to Immanuel Kant the historical trend has been to separate the logical content from the empirical aspects of geometry. The great importance of Euclid's *Elements* is that in it the theorems of geometry are proved without obvious reference to the world of experience. They are based entirely on logical deductions from a set of axioms.

But geometry has undergone a change since the fourth century B.C., first very slowly, and then quite rapidly. Its history may be represented, for our purposes, by the fate of Euclid's famous axiom of parallels. "And that, if a straight line falling on two straight lines makes the angles, internal and on the same side, less than two right angles, the two straight lines, being produced indefinitely, meet on the side on which are the angles less than two right angles."[2]

This axiom did not appear self-evident to many of Euclid's successors. Through the centuries numerous attempts were made to prove it from generally acceptable principles, but without success. Girolamo Saccheri, at the turn of the seventeenth century, was probably the first mathematician who began weighing alternative hypotheses and their possible consequences.

But non-Euclidean geometry did not begin until the work of Gauss. Although he published nothing on it, it became known after his death that he had worked out the foundations of what we now call hyperbolic geometry. This is one of the possible non-Euclidean geometries whose distinctness from Euclidean geometry turns upon its rejection of the postulate of parallels.

Ferdinand Karl Schweikart (1780–1859), a professor of law at the University of Marburg, is a much lesser-known figure in the history of geometry. In his student days he became sufficiently interested in the problem of parallel lines to have written and published a book about it in 1807. Though his approach was Euclidean, he arrived at the beginnings of hyperbolic geometry independently of Gauss and of the other originators of this discipline. In a letter to the astronomer Christian Ludwig Gerling, a student and friend of Gauss, Schweikart spoke of two distinct geometries. One, he wrote, is a geometry in the "narrower sense"; this is Euclidean. The other, the non-Euclidean version, he called "astral geometry." The implication was that the differences between Euclidean and non-Euclidean geometries should manifest themselves only at astronomical distances. Gerling communicated this letter to Gauss in January 1819. Gauss described his reaction to Schweikart's remarks as being very much after his own mind and, in rather enthusiastic words, began speculating about the implications of an *Astralgeometrie*.[3]

Gauss's immediate endorsement of Schweikart's distinctions foreshadowed the role that non-Euclidean geometry played in the general theory of relativity. The term itself is most felicitous and for that reason, in the proper context, I will refer to the geometry useful in general relativity theory as astral geometry.

It is possible that Gauss's response reflected his conviction that geometry should be regarded as a form of empirical knowledge. In a letter to the astronomer Olbers, dated April 28, 1817—two years before the Schweikart-Gerling-Gauss exchange—he wrote:

> I am becoming increasingly convinced that the necessary relations of our geometry cannot be proved, at least not by and for human reason. Perhaps in another life we shall attain such other insights into the nature of space as are now beyond our reach. Until then, we should class geometry not with arithmetic, which is purely a priori, but with mechanics. . . .[4]

In studying the properties of surfaces, Gauss made use of the concept of a geodesic, a line lying within a surface, being the shortest distance between two points in that surface. In a plane the geodesic is a straight line; on a sphere it is an arc of a great circle; on a doughnut, it is a variety of curves. Gauss showed that the nature of the geodesic depended on a property of the surface which he named its curvature. For a plane the curvature is zero, the radius of curvature is infinite; for a sphere the curvature is a fixed number, as is the radius of curvature; for a doughnut both variables assume a family of values.

It was the German mathematician Riemann who first generalized the concept of curvature to manifolds of any arbitrary dimensions by formulating the notion of multiply extended magnitudes. In 1854, in his formal, initial lecture at Göttingen he remarked that

> geometry presupposes not only the concept of space but also the first fundamental notions for constructions in space, as given in advance. . . . The relations of these presuppositions is left in the dark. . . . [This darkness] has been lifted neither by the mathematicians nor by the philosophers who have labored upon it. The reason for this lay perhaps in the fact that the general concept of multiply extended magnitudes, in which spatial magnitudes are to be comprehended, has not been elaborated at all. Accordingly, I have proposed to myself at first the problem of constructing the concept of multiply extended magnitudes out of general notions of quantity.[5]

He concluded that multiply extended magnitudes—today we call them n-dimensional manifolds—are susceptible to various metrizable relationships. Metrization means the possibility of assigning to a manifold a method whereby distances between any of its points may be measured. The metric is the criterion or formal instruction on how to metrize. It determines the size of the geometrical magnitudes such as lengths, areas, volumes, curvatures, or angles. The space of our kinesthetic experience is only one particular three-manifold. To learn about it—and to revert back to Riemann—there arises "the problem of searching out the simplest facts by which the metric relations of space can be determined. . . ." Once these are identified, the next task at hand will be to "decide concerning the admissibility of protracting [the metric qualities so identified] outside the limits of observation, not only toward the immeasurably large, but also toward the immeasurably small."[6]

We thus have a general notion of space, of Plato's home for all created things. We have some idea about how to assign topological properties to this home or, rather, homes. We know how to provide means whereby those topological properties may be measured and with Gauss and others, a desire to extend our findings to the geometry of the universe. These elements did not really come together until the work of Einstein.

For the moment let us backtrack to the invention of analytic geometry by Descartes and Fermat during the first half of the seventeenth century. The analytic use of coordinates made it possible to replace each point p in a plane by a pair of numbers in a tabulation. Each pair gives the location of the point with respect to two orthogonally intersecting lines whose orientation is assumed to be known. As a point moves, different pairs of numbers are assigned to its changing positions. The procedure reminds one of telling time by the conical dial of section 1.1. That task was also reduced from specifying directions and lengths of shadows to that of specifying numerals. Numbers can be more easily manipulated than shadows, directions, or distances. Indeed, if a point in a plane may be identified with two numbers, in a space with three numbers, why not imagine one of Riemann's n-dimensional manifolds in which points may be identified by 4, 5, 6, 10^3, or 10^{76} numbers? Such spaces may conceivably provide a system of versatile homes for more created things than Plato ever dreamt of in his philosophy.

Geodesics being the shortest distances or least curved lines between two points in a space, they may be spoken of, by analogy, as the straight lines of that space. Mathematicians located external to two spaces may then tell us that one of these "straight" lines is curved more than the other one, depending on the curvatures of the spaces. The degree of the curvatures is determined by the metrics of the spaces. The metric of "flat" Euclidean space permits the calculation of distances between two points from the coordinates of those two points, by means of the Pythagorean formula. The same formula remains valid for curved, non-Euclidean space, but the terms of the formula must be modified by the metric coefficient.

In ordinary three-geometry we speak of points, lines, two-dimensional surfaces, and three-dimensional volumes. Four-geometry permits one more degree of freedom. In a four-dimensional world there are points, world lines, two-dimensional surfaces, three-dimensional surfaces (often called hypersurfaces), and four-dimensional domains. A hypersurface may, for instance, be a spatial domain (a volume) at one instant of time. Or, it may be a two-dimensional surface for an extended period of time. Both of these hypersurfaces have three dimensions, embedded in a four-dimensional domain.

Surfaces or volumes in any geometry have intrinsic and extrinsic properties. For instance, the area of a spherical triangle depends on the excess of its three angles over 180°. This is an intrinsic property of the two-surface, the surface of the sphere. Spherelanders may discover and test this rule while remaining entirely within their two-dimensional world. An external property is something that demands a world extrinsic to a surface. For instance, casting a circular shadow upon a plane is an extrinsic property of the two-surface of the sphere. But mathematicians among the spherelanders could learn about the circular shadow by writing down construction rules for it.

Imagine that they have monitored the angles of a reference triangle and found the excess of its angles over 180° decreasing. They may then be able to derive a single parameter, a distance, a variable that they may call the radius of the world circle, whose changes can be related to the decreasing sum of the angles of the reference triangle. They may even conclude that the substratum of their universe is undergoing a change according to a stable rule. All they need to have is a theodolite, a knowledge of geometry, and a sense of time.

The purpose of this example was to demonstrate that with generalized ideas of space and appropriate mathematical machinery, certain new freedoms become available to those interested in the large-scale features of their universe.

6.3 The Astral Geometry of Einstein

We have seen how the special theory of relativity tied kinematics and electrodynamics to the absolute motion of light. Later we watched distances, times, causations, and particles emerge from the atemporal world through a process to which both the special theory of relativity and quantum theory bear witness. The general theory of relativity carries us a step further. It ties the effects of massive aggregates of matter to the absolute motion of light through a particular modification of the astral geometry of Gauss. In its Einsteinian form it becomes the geometry of a four-dimensional manifold in which distances, times, and massive matter appear as different properties of the same continuous field.

In a 1911 paper Einstein spoke about two systems. One is so oriented that the lines of force of a gravitational field run parallel with it and point in the negative Z direction, in the conventional triad of the Cartesian X-Y-Z coordinates. The other system is in free space, away from gravitational fields but is accelerated uniformly in the positive Z direction. No experiment performed on free material points in either system can reveal whether the system is the one in the gravitational field or the one being accelerated. On the ground of this thought experiment Einstein postulated the physical equivalence of the two systems. This assumption, later named the strong equivalence principle, "makes it impossible for us to speak about absolute acceleration of the system of references, just as the usual theory of relativity forbids us to talk about the absolute velocity of the system. . . ."[7]

Rather crudely: I can weigh a watermelon in either system by watching, for instance, the elongation of a spring. But I must decide from external evidence whether the lengthening of the spring is caused by the vicinity of a large mass, or by the acceleration of the support holding the spring, or by a combination of both.

Following Einstein's hint for the drawing of a parallel with the "usual theory of relativity," we note that the principle of relativity is itself an equivalence principle. It stands for the impossibility of distinguishing experimentally between relative translation with respect to an arbitrary body said to be in rest, and one which in fact is in absolute rest. Newton recognized that the distinction could not be experimentally demonstrated by observing "the position of bodies in our region" (sec. 3.1). The equivalence principle of special relativity postulates that it could not be demonstrated anywhere and, hence, an absolutely resting frame does not exist. One consequence is that we may now regard space and time as aspects of—as having emerged from—absolute motion, when subluminal speeds became possible. Consider next that subluminal speeds may only be reached by means of (negative) acceleration with respect to light. One consequence of the general relativistic equivalence principle is that we may now regard matter, once again, as having emerged from absolute motion when subluminal speeds became possible. The change from the Parmenedian universe of Newton to the Heraclitean universe of Einstein is now complete, having been accomplished through the application of two equivalence principles.

In both relativity theories light beams propagate along nul geodesics. The task of the Egyptian rope stretchers was taken over by light beamers. These people travel around the universe with powerful lasers, beaming them now this way, now that way. In whatever direction their light travels, the beams will trace the shortest possible distances between any two points along their paths, as did the ropes of the *harpedonaptai* for their universe. But the light paths, compared with what they would be in an empty universe, will be curved. The degree of their curvature will be determined by the metric of the four-dimensional domain in which they happen to propagate. Einstein taught us how to relate the metric to the amount and distribution of matter in the neighborhood of the light beam. Matter itself may thus be thought of in terms of a geometrical property of space-time, namely, its curvature.

Identifying matter with the curvature of space-time is an example of the way in which the use of the generalized umwelt principle is implicit in mathematical physics. The concept of matter derives from the classical, sensate description of the world; it is a representation eminently suitable for daily life and for most of science. The concept of space-time curvature comes from astral geometry. That matter is identical with the curvature of space-time is a report about the umwelt of the physical universe whose properties, as seen from within, are alien to our senses. We have no other access to it than through signs and symbols. In quantum theory the best we could do was to call an elementary object a propagating probability wave. Here we call matter geometry. Strange as the idea sounds, we must assign to it the same status of reality as we did to the probability wave. Doing this grants us a freedom of exploration which is not otherwise available. What follows in the balance of this section takes advantage of that freedom.

In special relativity theory, as a traveling clock approaches the speed of light its ticks become less frequent compared with a stay-home, inertial clock. After a burst of infinite proper acceleration, the moving clock enters the atemporal world.

Instead of letting the clock accelerate up to the speed of light, allow it to rest right here and begin the construction of a star, next door. By the equivalence principle of the general theory, the massive body should have the same effect on the clock as acceleration does. The observer may remain at a safe distance from the growing star. As that body increases its mass, general relativity predicts a slowing of the clock, a phenomenon that has been experimentally confirmed. At a certain level of the gravitational potential the clock will break into pieces. Its parts will become denser and denser and increasingly disorganized. When the curvature of space-time in the neighborhood of the (former) clock makes the path of a light beam reentrant, the clock will have stopped ticking because it will have been absorbed in the chaos of the black hole. Its functions will have become atemporal processes. The clock has descended along the scale of temporalities from the eo- to the proto- and finally into the atemporal world.

From clocks journeying in the galaxy we turn to galaxies journeying in the universe. Following E. A. Milne, galaxies are often described as the fundamental particles of the universe. Because the volume of the average galaxy is some fourteen orders of magnitudes below the estimated volume of the universe, the name is quite appropriate. Observation shows the universe to be isotropic; regardless of the direction in which one looks, the distribution of fundamental particles has the same density. Assuming we are not at a privileged location, it must be concluded that the universe is also homogeneous. In other words, the universe must look isotropic from any of its galaxies. It cannot have an edge, for instance, because from a galaxy situated along that edge, the universe would cease to be isotropic.

The primordial universe is believed to have possessed a vast but finite amount of energy: all the energy there is. The matter equivalent of that energy is all the matter that can be, eschewing the creation of energy ex nihilo. Therefore, one would expect the universe to remain finite in its matter and energy content and its size. But it must also be homogeneous. How is one to construct a universe that is finite but has no boundaries? By an appeal to Einstein's astral geometry.

Take four balls and call them galaxies or fundamental particles. Construct a system where they are mutually equidistant. A tetrahedron with a galaxy at each corner will satisfy the demand. This is your universe: it is homogeneous as far as its particles go but it is certainly not isotropic. If standing at one of the corners you look away from the three other galaxies, all you see is the void.

Now, construct a system with five mutually equidistant galaxies, using yardsticks.[8] It is impossible because we are still working with the local, "narrower" geometry of Schweikart. Change to Einstein's astral geometry and the construction becomes possible, provided that instead of rope-stretchers we employ light-beamers. They will measure the intergalactic distances along nul geodesics, following the prescription of Einstein, already employed when we used clocks to measure the lengths of moving yardsticks. Take next 10^{10} galaxies and employ general relativity theory for the construction of a world model. For a certain threshold value of total matter you have constructed a universe that is finite but unbounded. The model must remain an abstract one, described in the signs and symbols of mathematical physics; we have no direct experience of the cosmic umwelt. We could not have such experience.

The geometrical curvature of the universe in our epoch is not zero: that would make its astral geometry Euclidean. Neither is it infinite or even very large: that would make all of us into very dense radiation. It has a nondimensional value of around 4.85.[9] It has no regions that could be considered its boundaries. It is finite in its dimensions and matter content and possesses a scale factor, its curvature, in terms of which one may speak about its history—provided the speaker-geometer has a sense of time.

Let us catch the universe by its curvature as scale factor and accept an invitation for a journey to the big bang. Of course, we must talk the language appropriate to our own umwelt: long-term futures, long-term pasts, the arrow of time, a wealth of causations; all the paraphernalia of the noetic world. We use the same paraphernalia when we tell Fido to jump into the car, we are off for a two-week vacation to upstate New York. He need not understand the two-week vacation, all he needs to do is jump. With the universe it is worse but, fortunately, it is we who give an account of the world and not the other way around.

Since it is easier to think along the arrow of time than to think against it, we begin at the beginning.

There is the primordial fireball at time zero. Not really at time zero nor at volume zero because quantum mechanics forbids us to imagine, as physically meaningful, distances below about 10^{-33} cms. and times less than about 10^{-43} seconds.[10] (Earlier we spoke about the atomic chronon of 10^{-23} seconds as an atemporal boundary of the universe. Both the 10^{-43} and 10^{-23} second figures are derived quantities. The reader may have a choice.) The temperature of the fireball at that time zero, which is not exactly zero, is not necessarily infinite. D. C. Kelly has given some interesting reasons why adding energy to a gas around $10^{12} \,^\circ K$ will result in an increase in the number and variety of particles but not in an increase of temperature.[11] It is conceivable, therefore, that there exists an upper boundary to temperature just as there exists a lower boundary. Either conditions are imagined as approachable but not actually achievable by the temporal structures and processes of the physical world.

There are many conjectures regarding the composition of the universe close to the unreachable zero point of time. These conjectures generally hold that in the primordial state all constituents of the world were in thermal equilibrium. During the first ten seconds after the expansion began the density and temperature were so great and the interactions among atomic nuclei so fast that nuclear composition had to be determined by instantaneous conditions only. There could have been no correlations among any two states either locally or globally, there could be only pure chaos. Matter and antimatter are usually assumed to have been around in equal strengths because no one has thought of a good reason why they should not have been. But something in this primordial state, an asymmetry not yet recognized or perhaps pure chance, has led to the initial predominance of matter. Once assured, antimatter became metastable, if for no other reason than for the presence of an overwhelming amount of matter.

From about ten seconds to about one hundred seconds after the vast bang of the tiny universe, the world is believed to have consisted entirely of radiation in the form of light. The density of light had to be immense, perhaps of the order of 10^{93} gr/cm^3. In a radiative universe, with all objects traveling at the speed of light, no meaning could be given to a frequency scale; the world was atemporal. For a ghost traveling on a photon the universe had to look black. We may say, therefore, in the beginning was darkness.

The radiation-dominated era gave way to a matter-dominated universe perhaps at 10^{10} seconds—a thousand years—after the atemporal conditions had ceased. In another 10^8 years galaxies began to form and have remained with us since. During this period the average temperature of the universe dropped to around $3°K$, detectable as a relic of Creation in the microwave landscape of the cosmos.

All through this history except for the first 10^{-43} or 10^{-23} seconds, the astral geometry of Einstein remained a valid, formal representation of the physical world. It tells us that the universe is dynamic and undergoing a continuous change of its intrinsic geometry. Three-dimensional space has no extrinsic geometry because there is no space but space. However, working within three-space there does exist a variable in terms of which the dynamics of the universe may be understood. It is called cosmic time.

6.4 Cosmic Time

In experimental knowledge all findings must be testable by independent observations. One may drop weights from the Tower of Pisa or from an Atwood machine, in Massachusetts or in Kentucky; experimental answers to the same questions must be identical. This protocol is often interpreted as a call for collective approval, which it surely is. But more fundamentally, it is a necessary precondition for being able to separate in physical processes the contingent from the necessary or, as it is sometimes said, the boundary conditions from what is lawful and permanent. If we could perform only one single test upon freely falling bodies, there would be no way to determine whether velocity and acceleration depend on the objects dropped, on the air in which they fall, on the closeness of the State House or on the day of the week. To be able to separate the necessary from the contingent we need many tests, performed under diverse conditions.

But by definition there is only one universe. It is possible to play with words and fancy other universes of the same kind but then what is the name of the set of universes but a new, enlarged, but still single universe? The behavior of our world cannot be checked against tests performed upon other universes so as to determine which of the results are peculiar to an object called *our universe* and which are but contingencies dependent upon particular experimental conditions. Theories of physical cosmology must be judged, therefore, not only by the empirical confirmation of predictions based upon them, nor by their explanatory power alone (or together with confirmed predictions) but also by their general acceptability to critical thought. Accordingly, this section evaluates certain cosmic facts and cosmological statements in terms of the principle of temporal levels.

Expansion   What may be said about the universe with some degree of certainty?

The universe is isotropic: this is a matter of observation and has already been mentioned. We have also concluded that it must be homogeneous. These two features and the coextensive character of space, time, and matter are accommodated with remarkable elegance by general relativity theory.

The next fact to which most physicists would agree is the

cosmological Doppler effect or red shift. It signifies a systematic displacement toward longer wavelengths of lines in the spectra of distant galaxies, as well as a displacement of the continuous part of the spectrum. The red shifts of cosmological objects correlate with the apparent magnitudes of those objects in such a way that if the shifts are interpreted as Doppler effects, the recessional velocities are found to be proportional to distances, regardless of direction.[12]

The homogeneity discussed thus far was a static one. To it we must add a global, dynamic homogeneity, the idea that the cosmic change is uniform across the world. The systematic change observed from our galaxy is assumed to be the same as would be observed from any other fundamental particle, and would be found to follow identical laws. To those laws, if the dynamic homogeneity is correct, identical boundary conditions must apply, regardless of the position of the fundamental observer.

The only conclusion one can reach by considering these various details together is that some time in the past the immensity of our universe was smaller than it is today, as measured on a suitable scale. What we are doing, rather obviously, is tracing our path back to the big bang. Upon present evidence, following broadly accepted assumptions and using an appropriate general relativistic model of the universe, the age of the world in our epoch is calculated to be around 2×10^{10} years.

The reasoning just given is beneath the frequent earlier reference to the big bang and the expansion of the universe, and constitutes the origins of that idea in physics. The two concepts, the big bang and the expansion of the universe, have become household words in our epoch, but by no means matters of household understanding.

Cosmological common time   Distances within galaxies do not change as part of the cosmic expansion. It is only the atemporal and prototemporal strata that expand, containing electromagnetic radiation, gravitational radiation, neutrinos, and hydrogen. The cosmic motion carries with it the small eotemporal islands known as galaxies.

If all material in the universe were smoothed out, the result would be a cosmos of thin dust at near zero pressure. But the universe is not so constituted: it is organized, instead, along levels of different complexities. The galaxies themselves show a hierarchical grouping. Our own Milky Way forms the local group together with perhaps twenty other galaxies. Clusters of galaxies normally contain around 200 members, though some are known to contain as many as 10,000. It is suspected that clusters themselves group together into superclusters. Should cosmic dynamics so demand, there are enough galaxies to form extrasuper clusters.

Galaxies occupy an intermediate level in a hierarchy of self-gravitating systems. The spectrum of size extends from the sub-planetary masses just large enough to remain together to galactic superclusters. The structuring that has emerged from the unstructured primeval universe, in the words of David Layzer,

> has resulted from a continuing interaction between two opposing processes: hierarchical construction, which progressively generates new kinds of self-gravitating systems; and tidal disruptions, which tend to break down existing structure in gravitationally bound systems.[13]

A recent computer simulation involving 4,000 masses representing galaxies, moving under gravitational attraction in an expanding universe, showed a degree and kind of clustering that resembles the one actually observed.[14]

If the universe of two opposing processes is truly homogeneous and isotropic in its cosmic motion, it should be possible to identify an inertial frame, associated with each fundamental observer, such that in it the global expansion appears uniform. A frame of this kind will be at rest with respect to the average motion of matter in the galaxy because it is that average motion that represents the galaxy as a fundamental particle in the cosmic expansion. Physicists at rest in any of the fundamental coordinate systems should arrive at the same figure for the age of the universe.

According to the rules of time measurement, cosmic time will have been measured only after at least two fundamental clocks had been compared and the validity of the proposed transformation equation demonstrated. In this case, the equation is an identity. The task is very complex but present evidence suggests that the postulated identity corresponds to fact.[15] We assume, therefore, that a cosmological common time exists. The earth moves with our galaxy and, apart from local deviations, represents a privileged, fundamental frame. Our clocks measure cosmological common time.

General relativity has thus introduced the phenomenon of cosmic time into the expanding model of the universe.

Cosmological common time is not absolute time in the Newtonian sense because it is not independent of what happens in time. On the contrary, it is the cosmic motion that assures the universality of cosmic time. Furthermore, all clocks in motion with respect to the average motion of matter in the galaxy will deviate from cosmic time. But the reconciliation of cosmic time with special relativistic time is an intricate issue.

Let us remember the complete symmetry of time transformations in special relativity theory: all readings were mutually interchangeable among systems in uniform relative translation. It should also be recalled that the theory is adequate for the transformation of proper accelerations according to the principle of relativity. However, the theory could not deal with the behavior and effects of matter and, for that reason, the special relativistic universe remained essentially empty. I intend to show that the introduction of matter into the universe places certain boundary conditions upon the time transformations of the special theory and also leads, via the general theory, to the elucidation of certain asymmetries predicted by the special theory and known to be empirically correct.

Physical systems tend to seek out the lowest energy states available to them. Being at rest with respect to the average motion of matter in the galaxy, and hence in a frame whose proper time is cosmic time, represents such a lowest energy state. The rates of all clocks in relative translation with respect to the galactic clocks will be dilated by the special relativistic factor. None will go faster than the clock indicating cosmic time. Cosmological common time appears, therefore, as an upper limit to the rate of local time flow. Such a boundary is unpredictable from the special theory alone.

Next, let a clock be sent around a finite, unbounded but static universe. Let its readings be compared with those of a clock at rest with respect to the material background of the universe. This imaginary test, due to Whitrow, does not call for an assigning of time readings to moving clocks but envisages, instead, direct comparisons upon successive meetings. Whitrow has shown that the moving clock will record shorter intervals of time between successive epochs of meeting than the stationary clock would.[16]

The journey of the clock circumnavigating the universe does not involve nonuniform motion and, hence, on that account, no appeal need be made to the general theory. But the test itself pertains to a model of a matter-filled universe not imaginable within the domain of the special theory. Furthermore, if we wish to avoid completely the issue of assigning time readings to distant clock, then the two clocks must be synchronized side by side. After synchronization the transfer of the clock to the traveling frame must involve nonuniform motion and through it, a reference to the cosmos. The universe, eased out the door, came back through the window.

But the world of this thought-experiment does not have a cosmological common time because it is static. Let it expand, therefore, and create a cosmic scale.

In section 3.5 an asymmetry was noted between a resting and a traveling clock: the one at rest indicated 9,876 units of time between two events, the one on a round trip, 6,789 units between the same two events. It was decided that they both may be correct, provided the difference in their counts could be explained by a law of nature.

It was mentioned at the beginning of this subsection that galaxies do not themselves expand with the expanding universe. The dynamics of expansion and the phenomenon of cosmic time may be neglected if only local explications of physical processes are being sought. Accordingly, textbooks on special relativity theory routinely demonstrate that the smaller proper time of a traveling clock on a round trip, compared with the longer proper time of the stay-home clock, is associated with the curving world line of the traveler, that is, with nonuniform velocities. The instant-by-instant symmetry between the two clocks is maintained, however (sec. 3.5), and with it the complete motional symmetry predicated on the Lorentz transformations.

Let the experiment now be examined in cosmic terms. The accelerating phases of the motion will refer the time shifts to the material universe, via the principle of equivalence of the general theory. The expansion of the universe and the existence of cosmological common time cannot be neglected any more. From this viewpoint, clocks at rest in the fundamental frame of the galaxy represent privileged processes. Each of these clocks now corresponds to the stationary clock of the Whitrow thought-experiment even though it is stationary only with respect to the galaxy and accelerates with respect to all other galaxies. Each and every clock in motion with respect to the reference clock now corresponds to the moving clock of the thought-experiment even though its reference is a local framework.

The differential indications in the clock problem may thus be accounted for through the laws of local frame and a universe void of matter, if that is the perspective selected. The same phenomena may be described through the dynamics of the universe, if that is the perspective selected. We are witnessing a crisis of perspectives anticipated in Table 1 (sec. 2.2) and due to the fact that the concerns of the special and general relativity theories are two different integrative levels of nature. By the hierarchical, nested character of natural laws the general relativistic arguments subsume the special relativistic ones. Indeed, special relativity theory is satisfactory for the physics of Euclidean space-time, but it cannot be expanded to provide statements about the universe.

The cosmic dynamics of the universe may thus be seen as introducing coherence into an otherwise incoherent world of fundamental particles.[17]

The beginning of time   Before one can write any cosmology, scientific or narrative, one must assume that the universe does, indeed, function as a unit. Because of the way we understand enduring identities of all kinds, it is also necessary for all cosmological speculations to come to grips with the notion of a beginning or with the absence of a beginning.

At the turn of the fourth century Saint Augustine spoke for Christendom when he addressed an issue already ancient in his epoch. Eternity and time, he wrote, are rightly distinguished by this, that time does not exist without some movement and transition, while in eternity there is no change. But, he reasoned, there were no creatures around before Creation in whose movements time could pass, hence there was no time. There was only God in whose eternity there is no change at all, He himself being "the Creator and Ordainer of Time." It follows that "the world was made not in time, but simultaneously with time." Although He could have done so at any other time, we are not to "suppose that God was guided by chance when He created the world in that and no earlier time."[18]

Basic concerns related to a beginning of time have hardly changed in sixteen centuries. But the way the questions may now be asked are different and with that difference comes the possibility of new answers.

The mental image informing most inquiries about the beginning of the world is appropriate only for the biotemporal and noetic umwelts. "In the beginning God created heaven and earth" implies there was nothing before the instant of Creation but that there was something thereafter. Time was one of the aspects of the world which was not, and then was.

According to the principle of temporal levels, Creation was neither followed nor preceded by other instants, because the relationship future-past-present had no meaning in the atemporal, or even in the proto- or eotemporal worlds. The nanoseconds, seconds, and up to a hundred million years, to the epoch when the galaxies began to precipitate, the umwelt of the universe was prototemporal. Its conditions could not even define continuity, let alone relations among events corresponding to the notion of before and after. We can only say, those early instants were in a prototemporal, statistical way contiguous with the instant of Creation. We are free to describe these early periods as if they were intervals of a bio- or nootemporal umwelt, provided we realize that the description is only a translation of reality into visualizable terms. It may help to remember that the umwelt of the tick is limited almost entirely to the smell of butyric acid emanating from the skin glands of mammals. Imaginary sharing of the umwelt of the tick is difficult; imaginary sharing of the umwelt of the universe during its early minutes is more difficult. That is why we need the symbolic language of science.

There are some aspects of the principle of temporal levels that Saint Augustine might have found consistent with his own views. He did write, for instance, "It is in thee, my mind, that I measure time."[19] And man, even for this pre-Darwinian saint, did not appear on the scene until some time Saturday the week of Creation. And just "What kind of days these were is extremely difficult, or perhaps impossible for us to conceive, and much more to say!"[20] Thinking about the hierarchical emergence of temporalities may not have been so strange to him. But he surely would have been disturbed when told that eternity is not changeless rest but restless change, and that continuity, identity, and rest constitute the new, the emergent aspects of the universe.

Temporal horizons   If the global dynamics of the universe will be found to include supraluminal speeds, then for all fundamental observers there will exist a boundary beyond which no object may ever rise above the atemporal horizon. Such a boundary is called an event horizon.* If the global dynamics is one of decreas-

---

* The special relativistic objection to supraluminal speeds is that no motion can go beyond the velocity of light and carry information. According to the usual rejoinder in defense of event horizons, the special theory applies only to inertial frames and there is no such thing as an in-

ing expansion rate we have a particle horizon. This is the general relativistic analogue of the absolute elsewhere of the special theory. In the special theory of relativity objects emerge from the atemporal world as and when signals from them reach the here and now of the observer. In the general relativistic universe these times for distant galaxies are lengthened by the cosmic expansion. But sooner or later all objects beyond a particle horizon will rise out of the observer's atemporal umwelt. The time of their appearance depends only on distance and the rate of expansion.

In sum, universes whose rates of expansion increase possess event horizons; those whose rates of expansion decrease possess particle horizons. The latter is the case for the model we have been assuming as the one corresponding to the actual universe.

The beginning and ending of time   Whether the rate at which the universe is expanding will or will not slow down has been widely debated.[21] The answer depends on whether or not the total mass of the universe is sufficiently large to make it into that closed, finite but unbounded system that we have been assuming it to be. The presently known mass is insufficient to accomplish the closure; the search has been on for the missing mass. Let us assume that the needed matter will be found: perhaps in the form of intergalactic hydrogen, or black holes, or burnt-out supernova, or in the mass of the neutrino, or in a combination of some of these. If it is, then the dynamics of general relativity demands a collapsing phase. The collapse is imagined as an inverted playback of the expansion of the cosmic motion. It would be detectable by each fundamental observer as a blue shift. It would happen at an accelerating rate until all energy will have regathered itself into the primordial atemporal chaos in a final big crunch, to use J. A. Wheeler's descriptive phrase.

The immense cosmic exhalation and inhalation must be assumed to remain a physical process, subject to the laws of physics; there are no compelling reasons to assume it otherwise. Except for cosmic thermodynamics yet to be considered (sec. 6.6) all physical processes considered thus far turned out to have been undirected.

---

ertial frame common to the universe. Although this is true, the defense is artificial because temporal structures do carry information. The difficulties related to supraluminal speeds are some of the reasons why models that demand it are intellectually unattractive.

The all-encompassing cosmic motion seems to be directed, from bang to crunch. Could it be the physical source of the biotemporal and noetic arrows of time?

In an examination of the topology of space-time Michael Heller concluded that a manifold is a suitable arena for physical process if and only if any of its points distinguishes between two time directions. "The result seems to be philosophically deep and interesting: the existence of a kind of local time—yet arrowless—is a precondition for every physics." From "the methodological point of view predictions are exactly as good as retrodictions, only psychologically are predictions of higher value for physicists." For the arrow of time, he concludes, we must look elsewhere.[22]

Heller's reasoning refers to local conditions: to "any point" in the universe. Is the expansion-contraction of the universe a global motion prevailing over all local two-wayness of time? Is it the irreversible process par excellence for which people have been looking?

Let us recall the definition of irreversibility in section 5.4. "An irreversible process is one for which nature has such a preference that the reverse process is meaningless." The reverse of the expansion is contraction, the reverse of contraction is expansion. We have been considering them liberally. They are certainly not meaningless. Or, taking a step further back, let us remember Planck's comments which inspired the definition. It is not sufficient to define an irreversible process as one that cannot run backward because although a process may not normally run backward, the initial conditions could, nevertheless, be somehow restored. In the case of the universe nature itself is working hard on the restoration of the primordial fireball, or at least so we have assumed. We conclude that the cosmic motion is not irreversible.

The red shift is certainly not arguable and the simplest explanation for it, consistent with general relativity theory and with our understanding of order in nature, is one of expansion. Consider, however, that the spectral shift of galaxies, appropriate for a universe expanding from past to present to future is identical with the spectral shift expected from a universe collapsing from future to present to past.[23] We could place either the time-forward or the time-backward interpretation upon the red shift. There is a good reason for selecting the time-forward mode: our sense of time. But then the shift itself cannot be used to justify the choice. Similar arguments can be advanced for a blue shift.

We have learned that the initial cosmic singularity—the bang —had no instants following or preceding it but only instants contiguous with it. Identical arguments must hold for the crunch. In fact, the same argument also holds for the whole eotemporal history of the general relativistic universe: it is one of pure succession from bang to crunch or crunch to bang.

The conclusion is curious but self-consistent: from within the eotemporal umwelt of the physical universe, for the model we have been assuming as valid, the beginning and ending cannot be told apart. As living and thinking beings we can distinguish between them because one is in the past and one is in the future, with respect to our present. But these are biotemporal and noetic distinctions. From the point of view of the universe itself and for want of a better word, the two events should be described by a single word, *begendings*.

The situation calls to mind the suggestion made by a number of authors, that the topology of cosmic time is that of a circle. Davies concludes his *The Physics of Time Asymmetry* by endorsing this "exotic possibility" because it leads to a description in which "good agreement with the observed universe is obtained."[24]

The idea is that at the end of a cosmic cycle time reenters itself but does so without implying a recurrent world, with everything in it done over and over again an infinite number of times. In other words, cosmic time with the topology of the circle remains the directed time of the biotemporal and noetic worlds, yet it is not a directed time because the future becomes the past. The result is causal anarchy.

The anarchy may be avoided if, instead of appealing to the topology of the circle, we remain with the topology of the undirected line. Cosmological common time, then, is the pure succession of kinematics and dynamics, the garden-variety *t,* the kind of two-way time with which physicists have learned to live quite well.

The stunning consequence of the eotemporality of cosmic time is that from within the physical universe, its two "begendings" become indistinguishable. The reason is not that directed time somehow reentered itself. Rather, there are no physical processes whereby the two begendings could be told apart. But then, by the identity of indistinguishables we are compelled to conclude that we have been contemplating only one single event. It is that of global creation and destruction, continuously present "beneath" our worlds as it were, but seen in two different perspectives from the nootemporal integrative level.

To try to elucidate this curious state of affairs, it may be well to remember an old truth. If someone claimed that the world was created only briefly before my birth, including all its living and inorganic furnishings, I could not prove him wrong. Neither could I demonstrate the untruth of a prophecy, concerning the end of the world two hours after I die. Conjectures regarding cosmic creation and destruction are subject only to tests of consistency with whatever else we know about the universe. This does not make time unreal. It only places cosmological common time in its proper position along the hierarchical organization of nature.

It is not by chance that the cosmic dimensions of time have awed and fascinated the intellects of all epochs. Often they have found their most articulate expressions in poetic metaphor. In T. S. Eliot's "Burnt Norton" we read,

> Or say that the end precedes the beginning,
> And the end and the beginning were always there
> Before the beginning and after the end.
> And all is always now.

So at least, for the three integrative levels of the physical world.

Black holes   General relativity theory added a new kind of atemporal boundary of the universe, to those already discussed: the surfaces of black holes. Much has been written about black holes and there is no reason to believe that the basic reasoning is wrong. But at this writing, black holes are still only postulated objects. They constitute one possible final stage in stellar evolution. In the words of Blanford and Thorne, in spite of much observational data across the electromagnetic spectrum, "astrophysics has still not advanced to the stage at which we can be strongly confident that black holes exist at all."[25] We will proceed in the belief that, like the missing mass of the universe, the existence of black holes will be experimentally verified.

Though astronomical bodies may endure for very long periods of time, they are basically unstable. They last only as long as the gravitational interaction among their particles—tending to compress the body—remains in equilibrium with internal pressures, forcing the body apart. If the internal, centrifugal pressure decreases below the centripetal pull of gravity, the object begins to collapse. As its density increases, the mass-equivalent of the gravitational pressure itself adds to the gravitational pull of matter and accelerates the collapse. When a certain density is reached the space-time curvature of the region becomes sufficiently large to make light beams re-entrant. Thereafter nothing may leave the region, not even light. To outside observers the object will appear as the absence of all things, a black hole.

As seen from a distance, the boundaries of a black hole form an event horizon: no object from beyond it can ever reach the outside. But objects getting sufficiently close to it may fall into it, increasing its size. It has been shown by theoretical arguments, supported by convincing analogies, that the areas of the event horizons determined by black holes are proportional to their entropies.[26]

Black holes can have mass, angular momentum, and charge, amounting to the total mass, angular momenta, and charge that fell into them. But all other structural or functional information gets degraded and completely lost.

According to theory, black holes vary in size from 10 to 20 kms. down to 10^{-12} cms. across. The microscopic black holes are imagined as having been around since the early days of the universe. Even the small ones may weigh as much as a billion tons each, at equally impressive densities. The internal temperatures of black holes are inversely proportional to their sizes. When they are large, their temperatures may be as low as 10^{-7}°K, thus approaching absolute zero. Microscopic black holes may be as hot as 10^{11}°K, approaching the limiting temperature of the big bang.

Although the closedness of space-time prohibits anything from leaving a black hole, from a distance it may appear as if it were emitting particles as any other thermal body would. The machinery involved is pair formation. It is imagined that pairs forming near to but on this side of our event horizon, separate. One particle falls into the black hole, the other takes off to announce the loss of its partner. This handsome mixing of general relativity and quantum theory is known as the Hawking evaporation. Since it takes place along an interface between an atemporal and a prototemporal world, it is surely a process also identifiable in the big bang.

Black holes, it is said, have no memory. The claim refers to the fact already mentioned: except for masses, charges, and angular moments, all structural and functional information is lost in a black hole. Furthermore, because of the tremendous forces at play, and because of the short distances involved, a black hole can support only atemporal elementary objects. Black holes are mini-universes within our maxi-universe, ready to do their mini–big bangs, should there be a reason for it.

The entropy of the black hole, a variable assumed to be proportional to its surface area, is an information type entropy. It relates to the amount of information that either did or could have gotten lost in it. The larger the black hole, the more there is in it to get disorganized, the larger the entropy. When two black holes merge they create a larger common surface and represent, therefore, a larger entropy. Some claim that through this behavior they follow the prescriptions of the second law of thermodynamics and show a direction of time. But this view is not shared by all. Hawking, for instance, maintains that black holes behave in a time-symmetric way.[27]

Black holes are examples of the atemporal boundaries to the universe; specifications holding for all similar interfaces hold for them as well. Thus, one cannot enter them without becoming chaotic; the behavior of matter, energy, and entropy along their surfaces is controlled by the laws of quantum theory; also, anything may issue from them as long as it behaves in ways appropriate to the boundary conditions. Since no system may be more disorganized than chaos, and because there can be nothing more atemporal than the atemporal world of physics, it makes no sense to talk about anything "beneath" a black hole. One cannot fall through a black hole on the way to another universe.[28]

The two world systems Newtonian mechanics may be stated in a form that makes gravitation a geometric feature, determined by field equations almost identical with those of the general theory of relativity.[29] Significant distinctions remain, however, in the conditions that must be imposed upon the solutions. In the case of the Newtonian universe, signals may propagate at infinite speeds, distances may be measured with rigid rods, clocks may measure absolute time. In the case of the Einstein universe, local conditions must be special relativistic, the motion of light has to become absolute motion and distances and times must separate out of the motion of light for subluminal velocities.

The major features of the expanding universe may also be treated in a Newtonian theory; the key equations of relativistic cosmology can be derived from classical mechanics.[30] The world model appropriate to the task is a smoothed-out dust universe, a continuous medium of low density. Light is assumed to propagate rectilinearly and at constant local velocity. What one obtains is an exploding gas for which Hubble's law holds: each fundamental observer sees all others receding from him at an accelerating rate proportional to distance. But since Newtonian physics cannot produce a mechanism to slow the expansion, this Newtonian model of the universe expands for ever.

A fundamental observer in it, noting the cosmological Doppler shift, must conclude that some time earlier all the matter of the universe was collected in a single region of his world. If he shies away from infinite densities he might assume his universe to contain only a finite amount of mass. If he is also a good post-Copernican he will refuse to regard his galaxy as privileged and postulate his universe to be homogeneous. By the arguments of section 6.3 he will construct a non-Euclidean, finite but unbounded cosmos. As in the Einstein model, he will also measure cosmic time by clocks attached to fundamental observers and his reference epoch will be that of Creation. He will have given up his idea of absolute space but not that of absolute time.

Two essential differences will remain, distinguishing his from the general relativistic universe. The first one is that in the Newtonian universe the expansion can only be an empirical fact without theoretical justification, whereas in the relativistic model it follows from theory. The second difference is more subtle.

The time of the Newtonian universe acquires its arrow unproblematically: it is given. It represents the absolute flow of directed time, the passing of the created in the mind of the Creator. The physicist is privileged to be aware of this flow because he shares, ever so slightly, the mind of God. The act of Creation was a manifestation of simultaneities of purpose, to wit, the purpose of commencing history. If the t of the Newtonian model does not respond to the direction of time it matters little: as everyone knows, time flows from past, to present, to future, even if it is admitted that the phrase, "the flow of time," is metaphorical.

The Einstein model is the product of a very different world view. In it, everything observable must be accounted for through the behavior of nonliving, nonthinking matter. Starting with these, usually hidden, precepts the theory succeeded in constructing a cosmology which can describe the global peculiarities of the dynamic, finite, but unbounded universe correctly and self-consistently. It can and does dispense with absolute time in favor of cosmological common time. And, as we have seen, this cosmic time is one of pure succession. The very success of relativistic cosmology demonstrates, therefore, that the lifeless and mindless universe of physics is eotemporal.

6.5 A Glance at Unified Fields

Whether one claims that matter is made of space-time or that space-time is an extension of matter is only a question of arbitrary focus, not in itself guided by the theory. But whichever way the statement runs, the idea of mechanical forces acting upon charged and magnetized particles has to be retained. The field equations of general relativity theory break down in certain regions of space-time; they form mathematical singularities that are taken to represent particles.

Because of this mathematical fly in the physical ointment, much has been written about the desirability of a unified field theory. One envisages a formalism broad enough to permit the description of the particles themselves as general relativistic modifications in the geometry of space-time. Einstein spent many years of his life trying to obtain a suitable formulation but he did not succeed. Many great physicists have applied their talents to the question of how to extend space-time descriptions to the microscopic world. But their work has shown thus far only peripheral success (see sec. 2.2). Perhaps all we need is more patience, time, money, and energy, for the task is prohibitively difficult. Alternately, perhaps the work is based on certain unstated assumptions that do not correspond to the true nature of the physical universe.

In the literature of time in physics, time has been assumed to constitute a single aspect of cosmic order from photons to compounds to animals and man, although, in Whitrow's words, "certain aspects of it become increasingly significant the more complex the nature of the particular object or system studied" (sec. 2.1). Many writers, though, have made allowances for the atomic world to be timeless.

What has been said thus far suggests that the temporality of the macroscopic order of matter is qualitatively different from the temporality of the microscopic order and both differ from the temporality of the radiative universe. It would follow that only such a formalism as allows for and accommodates these different temporalities can unify general relativity and quantum theory.

6.6 Two Cosmic Arrows

In a universe of finite material content the thermodynamics of inorganic evolution must recognize a monotonic increase of entropy, though not necessarily one at uniform rate—if measured in terms of an evenly flowing time. A corollary of this theory is a steady though not necessarily uniformly increasing state of ordering as one looks back at the big bang. P. C. W. Davies sums up the situation as follows:

> To explain where the ultimate cosmic order came from, and hence to account for [the] distinction between past and future, it is necessary to consider the creation of the universe—the big bang. The cosmic structure which emerged from the primeval furnace was highly ordered, and all the subsequent action of the universe has been to spend this order and dissipate it away.[31]

According to this argument, the origin of time's arrow must be found in the initial conditions of the cosmos.

At variance with this view, section 5.2 gave reasons why the act of cosmic creation must be regarded, instead, as simultaneously possessing the properties of ordering and disordering. It is the first instant to which the dynamic condition of coexistent organization and disorganization applies. In this view, what was created was not the possibility of dissipating a store of order until the end of the world, but the possibility of opposing order to disorder.

In a series of papers David Layzer put forth yet a different theory. He rejected the view that the universe needed to be highly ordered in its original state and maintained, instead, that the primeval condition was one of an undifferentiated state and without any structure.[32] He sees the expansion of the universe as generating information as well as entropy, order as well as disorder. More specifically, information is generated whenever the expansion (or contraction) rate of the cosmic substratum exceeds the rate at which local processes are able to maintain equilibrium. He singles out nucleogenesis at the early stages of the expansion—in our terms, the creation of the prototemporal from the atemporal world—as an example of a decreasing entropy process that has been thoroughly investigated. But this is only a single instant. The history of the universe is one of continuous ordering manifest in the accumulation of macroscopic structures. Layzer concludes one of his remarkable papers as follows.

> Although the continual generation of new information is characteristic of both biological and cosmic processes, the underlying causes are completely different. Cosmic evolution is a consequence of the cosmic expansion; biological evolution and the phenomenon of life depend on the interaction between biological systems and their environment—in particular, on the ability of biological systems to extract information from their environment. That information is there, however, because it was generated by cosmic evolution.[33]

Section 7.1 will give reasons why living organisms must be regarded as doing substantially more than gathering information: they generate a new kind of information which could not be generated by inorganic systems. But it does remain true that biogenesis and organic evolution could come about only because the earth did reach a certain level of organizational complexity. In its turn, this was made possible by the information-generating processes of the inorganic world with their ultimate root in the expansion of the universe.

Layzer identifies the origins of the arrow of time with the information- or negentropy-creating capacity of the universe, a process whose origins he has traced and discovered: "Cosmic evolution generates a uniquely defined temporal sequence of configurations of increasing complexity."[34] The arrow we experience points in the direction of that complexification.

This, however, is as hasty a conclusion to reach as is its opposite, the claim that time's experiential arrow points along the direction of decay. These arrows point in opposite directions, they define each other and have been doing so ever since the conditions for their coexistence were created. No physical or life process is conceivable without being governed by principles of ordering as well as disordering although locally these two need not, and usually do not, continually unmake each other. But the universe at large remains as Shakespeare's seashore, an "interchange of state," in which he has

> seen the hungry ocean gain
> Advantage on the kingdom of the shore,
> And the firm soil win of the watery main,
> Increasing store with loss and loss with store.

The universe at large is eotemporal, even from the thermodynamic point of view.

This discovery places the tests of section 5.4 on reversibility and irreversibility in a new perspective. The three physical tests could not have led to the identification of a preferred temporal direction because the entropy increasing and decreasing mechanisms they mobilized were only part and parcel of the two cosmic processes of growth and decay. They are, in Layzer's words quoted in the preceding section, the progressive generation of "new kinds of self-gravitating systems [and the] tidal disruptions which tend to break down existing structures in gravitationally bound systems."

We conclude that both on the cosmic and on the local scale, the physical universe remains eotemporal. For the sources of our experience of future, past, and present we must look elsewhere.

Some time ago, while trying to delineate the boundaries of the notion of time, I wrote that, through the life of all men and women

> personal identity becomes intelligible and communicable to others because of the existence of a subtle private and communal understanding of an ordering principle. The same understanding or type of knowledge, is also essential for the description of man's relation to the universe. This principle, or knowledge [is usually] couched in terms of an idea called time. . . .[35]

The ordering principle is the directed time of the biotemporal, nootemporal, and sociotemporal integrative levels of nature. The dynamics of these organizational levels are permitted, but not prescribed or demonstrated by the physical universe. For the head of time's fleeting arrow we must look elsewhere.

But the epistemological commitment of the hierarchical theory of time prohibits us from assigning the responsibility for our experience of temporal passage to the higher integrative levels, simply by fiat. Reasons must be given in support of the claim that future, past, and present are symptoms or correlates of the biological, mental, and social functions of living organisms.

7

Biogenesis and
Organic Evolution

The Breaking of the Eotemporal Symmetry

This chapter identifies the coming about of life with the breaking of the eotemporal symmetry of the physical world. It traces certain continuities, and notes certain discontinuities, between the inorganic and organic worlds, by focusing on issues of the life process rather than on those of living structures. The emergence of man is seen as the next step after biogenesis in the evolution of time, because it extended the boundaries of the physiological present by means of the mental faculties of humans.

The path that may thus be traced from the time-related teachings of physics to those of biology, psychology, and social science permits new insights into the usefulness of quantitative knowledge, the evolution of space, and the nature of scientific law.

7.1 From Simultaneities of Chance
to Simultaneities of Need

Scientific interest in the origins of life has been focused almost entirely upon issues of structure and chemical composition. As in physics so in biology and biochemistry, a tradition favoring the spatial has been evident. The practice and idea of dissection represents very well the preferred experimental methodologies and theoretical stances. We will approach biogenesis by a different route; it may be called, *timesection.*

Although awareness of cyclic behavior as a fact of life is surely as old as civilization, systematic exploration of biological rhythms is a very new interest. It did not begin until the pioneering work of Erwin Bünning in the 1930s.

Since its inception, the scientific literature of biological rhythms has been assigning priority to the discovery of the mechanism of hypothetical objects named biological clocks.[1] But an appreciation of the fundamental importance of physiological clocks for the phenomenon of life itself is only now coming into sight. Indeed, the significance of biological rhythm goes much beyond what even its most enthusiastic explorers have been claiming for it, because biological clocks constitute the obvious link between the inorganic and organic functions of matter. Therefore, an understanding of the nature of biological rhythm should lead to a better understanding of the coming about of life itself.

When trying to understand the work of the biological clock, biologists often use a particular metaphor. They talk about removing the hands of the clock, the external indicators of time, so that the underlying mechanism may be revealed. Everyone knows, of course, that the expression is a metaphor and no one is looking for clock hands. But metaphors are important because they represent nonmetaphorical beliefs which may or may not be correct and because they guide and direct our thought. In this case, the belief is that it makes sense to separate some things called indicators from other things called clocks. But, as we learned in section 2.1, time measurement demands a comparison between at least two processes, each of which must exhibit identifiable changes of state, called instants. It is not possible to have a clock without hands that would still indicate instants.

Begin with a regular alarm clock and remove its hands. You are left with two concentric shafts rotating at different rates. These are the new hands. Remove the shaft. You are left with rotating gears as the even newer hands. Remove the gears: there is no clock left.

When it comes to biological clocks, one may examine the cyclic behavior of organisms, level by organizational level. Beneath the cyclic functions of the organism as a unit there are the cyclicities of the cells and, beneath those, the rhythms of the organelles. Far beneath them come the oscillations of the molecules and finally those of the probability waves. The cyclic behavior of all organisms involve the collective, orchestrated temporal programs of all these processes together. What on one level may be regarded as the indicator of instants is, seen from the next higher level, the process itself. Remove the last indicators and you have destroyed the last process itself.

Working our way upward, it is clear that the simplest, the earliest living indicator of time is that inorganic process which forms the bridge between the inanimate and organic forms of matter.

B. C. Goodwin and J. T. Bonner are two among the many biologists, Jürgen Aschoff and A. T. Winfree[2] are two among the many physiologists who regard oscillatory behavior as the fundamental process of living matter. We already know that cyclic behavior is also characteristic of matter from electrons, to molecules, to planets. We should seek, therefore, a transition between the inorganic and organic worlds by taking advantage of this common feature.

When speculating about the origins of life, Darwin imagined "some warm, little pond, with all sorts of ammonia, phosphoric salts, light, heat, electricity" in it.[3] Let some such mixture be the primordial slime together with solid structures of various kinds, placed in the dynamical environment of the early earth. Then let us try to think of mesoforms whose functions may be shared by the most organized inorganic and the least advanced organic systems.

Needham noted that between the living and nonliving realms the crystalline represents the highest degree of organization.[4] J. D. Bernal spoke of generalized crystallography as the key to molecular biology and suggested clays as providing a suitable matrix for the synthesis of macromolecules.[5] The idea that living cells are actually liquid crystals has been receiving increased attention.[6] Following the clay-line reasoning, Cairns-Smith has put forward a theory according to which the ancestors of life were, in fact, crystals, made up of structures standing in for the DNA-RNA-protein system of modern biochemistry.[7]

Let us hypothesize that the biogenetic landscape did include crystals, occupying the protective, microscopic niches in existent geological forms, just as blue-green algae colonize air spaces today in the rocks of the Dry Valleys of the Antarctic. Based on a mixture of evidence and speculation recorded in the literature, let us assume an energetic environment including optical radiation, high frequency solar radiation, heat, electric discharge, and supersonic shock waves.

Let us now detour to a discussion of the thermodynamic corollaries of biogenesis.

According to C. E. Muses, beneath the many action laws of physics, such as Hamilton's principle of least action, LeChatelier's law of displacement against stress, or Fermat's principle of minimal paths, lies the unifying principle of minimal entropy increase.[8] Ilyia Prigogine in his earlier works reasoned that living organisms in their mature states near equilibrium are characterized by minimum rates of entropy production.[9] This led others to suspect that minimizing entropy production may correspond to certain policies of energy exchange, designed to produce the greatest economy of metabolism per unit weight. This particular conjecture was successfully tested for bacterial metabolism.[10] But living organisms, as we know them today, function far from equilibrium conditions. For this reason Prigogine and others attempted to extend the validity of the minimal entropy production principle to systems far from equilibrium. Prigogine himself summed up the situation as follows.

> The theory of minimum entropy production expresses a kind of "inertial" property of nonequilibrium systems. When given boundary conditions prevent the system from reaching thermodynamic equilibrium (i.e., zero entropy production) the system settles down in the state of "least dissipation."
>
> It was clear when this theorem was first formulated that it was strictly valid only in the neighborhood of equilibrium, and for many years great efforts were made to extend this theorem to systems farther from equilibrium. It came as a great surprise when it was shown that in systems far from equilibrium the thermodynamic behavior could be quite different—in fact, even directly opposite to that predicted by the theorem of minimum entropy production.[11]

The new direction of inquiry is the right one if one is interested in the entropic behavior of organisms advanced beyond the most primitive forms of life. But primordial organisms had to be close in their functions and structures to their nonliving cousins. If one's interest is biogenesis, it will still be necessary to remain with the principle of minimal entropy production. Later evolutionary development may then be followed up along whatever path development did in fact take, and handled by means appropriate for living systems increasingly distant from equilibrium.

Imagine a simple system of molecular oscillators operating very close to equilibrium and governed by the principle of minimal entropy production. The system may be thought to manifest the combined presence of two trends: a decay trend which increases entropy and a growth trend which decreases it. The best an inorganic system can do and still undergo change is to produce entropy at a minimal rate: the decay arrow will remain minimally longer than the growth arrow.

All principles governing the processes here envisaged are statistical. An occasional crossing over from minimal entropy production to the generation of information is, therefore, a matter of chance. If suitable machinery is available, a new system of information generating molecular oscillators becomes conceivable.

We have identified continuity as well as discontinuity between a group of inorganic and a group of proto-organic molecular oscillators, operating near equilibrium. The continuity is the recognition of the growth and decay arrows in both systems. A search for the origins of life should involve, therefore, the identification of such mesoforms as seem suitable to make the transition from a local growth arrow shorter than the decay arrow, to a local growth arrow longer than it. The discontinuity is the need for the proto-organic molecular system for the maintenance of its internal coherence, if the system is to remain at least partly autonomous.

Perhaps some of the crystals of the biogenetic landscape were able to absorb oscillating energy at certain frequencies and use it to maintain oscillations at such other frequencies as fitted the frequency spectrum of their geological niches. Few of our environmental periodicities today are rapid enough to approximate the natural frequencies of inorganic crystals, but our environment is substantially different from that of the Cryptozoic era. One may imagine that those crystals whose natural frequencies did fit the dynamics of their niches had a better chance for maintaining their structural identities than those whose natural frequencies did not. What all this amounted to—if indeed it happened—was the copying of environmental periodicities, with a statistical spread. The spread made it possible for the environment to exert selection pressure upon the available populace. One may then hypothesize that natural selection thereby supplanted chance, as simultaneities of need supplanted simultaneities of chance.

The epoch here imagined, the transient period when order in structure came to be supplemented by temporal ordering in function, is the true origin of life. It is the least explored era of biogenesis. The hypercyclic system of Manfred Eigen[12] is applicable only to chemical structures whose complexity is much above what the complexity of the first living clockshop is likely to have been.

A system of coupled oscillators can endure as a unit only if, and only as long as, its individual clocks remain mutually supporting rather than destructive in their physical and chemical behavior. In different words, their cyclicities must be coordinated. Therefore, the logistics of the small clockshop had to include such provisions as were necessary to insure that certain changes do, and some others do not, happen simultaneously. Otherwise the collective viability of the system was jeopardized. It is this necessity for internal coordination that constituted the birth of presentness. Though the eotemporal symmetry was now broken, life retained a continuity with both the growth and the decay arrows of the physical world.

However, while the principles governing the decay processes are shared between organic and inorganic systems, those which generate information are distinct. The information creating processes of the universe are global and involve simultaneities of chance. Those of living organisms are local and involve simultaneities of need, defined instant by instant. Also, while inorganic evolution is open-ended, it seems to have certain boundaries of complexity beyond which it does not go. Organic evolution is also open-ended but seems to have no boundaries as far as the variety of living forms can go, though it seems to have a boundary as far as complexification goes (sec. 7.4).

A recent paper by M. C. Moore-Ede and F. M. Sulzman on "Internal Temporal Order" begins:

> The temporal organization of physiological events within an animal may often be as important as their spatial organization. Mutually interdependent events must not only occur at precise spatial locations but must also occur with appropriate timing. Similarly, incompatible processes which require different physico-chemical conditions for their completion, can be separated just as effectively in time as in space.[13]

In the radical process philosophy of the principle of temporal levels, the preceding quote would acquire a different emphasis.

The viability of temporal organisms depends unequivocally upon their temporal organization. Mutually interdependent events must occur with appropriate timing as well as at appropriate spatial locations. Incompatible processes must be separated temporally, otherwise they lead to the decay and eventual death of the organism. Therefore, for the living organism certain events must and others must not be simultaneous, if it is to maintain its integrity.

Figure 12 illustrates the time-related issue involved.

Figure 12. Biogenesis broke the temporal symmetry of the physical world by creating the conditions necessary for a functional definition of a present, to wit, the need for internal coordination. Future and past could thus acquire meaning with reference to the physiological present. In the picture, the time of the birthday cake has no arrow; that of the birthday dragon, does. Courtesy, Recycled Paper Products, Inc. All rights reserved. Design by Sandra Boynton. Reprinted by permission.

In several lectures, John von Neumann sketched the specifications an automaton must have if it is to be self-reproducing. He appealed to the idea of complexity which he proposed to relate to the "crudest possible standards, the number of elementary parts." Further, he wrote:

> There is this completely decisive property of complexity, that there exists a critical size below which the process of synthesis is degenerative, but above which the phenomenon of synthesis, if properly arranged, can become explosive. . . . [Each automaton that has passed the crucial threshold of complexity, will be able to produce other automata] which are more complex and of higher potentialities than itself.[14]

If this rule also holds for living matter, then one would expect reproductive capacities to be of more recent origin than life itself. This is believed to be the case. Only after the small clockshops reached a certain level of complexity was it advantageous to reproduce through heirs that develop rather than by replicating identical heirs, because only then would the offspring be better adapted than the parent. The evolutionary precedence of living functions to reproductive ones suggests a period in early organic evolution when the umwelt of living matter remained almost purely cyclic—almost eotemporal—with a broad and ill-defined present. As the internal coordination of living organisms became increasingly refined, the boundaries of the present began to narrow.

In the model of biogenesis here proposed, improved fitness consists of improved copying of the spectrum of environmental cycles. By increasingly finer tuning, each new clock, when integrated with the older ones, provided an increased measure of adaptation, increasing the viability of the clockshop as an autonomous unit.[15] We may imagine the biological clocks of the proto-organisms as gradually filling the available niches in the rhythmic spectrum of the geological, astronomical, and climatic environment.

The characteristic variable of such a model is complexity. Thermodynamically, complexity relates to information and entropy: the more complex a system the more information is necessary to specify the number of states in which it might be found. Also, the more complex it is the more there is in it to get disorganized, hence its entropy content may be greater. Evolutionary fitness then becomes analogous to potential energy. Based on such premises, Saunders and Ho formulated a principle of minimum increase in complexity, analogous to the principle of minimum entropy production by Prigogine.[16] From what we have learned, the principle of minimal increase in complexity should also apply mostly to systems near the thermodynamic equilibrium of inorganic matter, while systems far from equilibrium remain free from such restrictions.

Whatever the details of evolutionary development might have been, internal complexification had to lead, eventually, to the generation of internal cycles to which nothing in the external world needed to correspond.[17] Some of these might have been necessary to maintain the internal viability of the organisms. Henceforth, natural selection acting upon externally manifest rhythms (paradigms of phenotypal behavior) came to effect internal rhythms (the ancestors of genotypal behavior). Under continued environmental pressure by a spectrum of external rhythms, one may imagine the evolution of the cyclic order of life. The adaptive process that Eigen and Winkler-Oswatitch[18] have perceived in the behavior of transfer-RNA molecules, could have become the controlling mechanism of organic evolution only after, what Eigen has called, molecular quasi-species have come into being.

With life born as a miniature shop of coordinated clocks, which way did life evolve? It expanded the spectrum of rhythms.

The cyclic spectrum of biological clocks in the species alive today is very broad. Human skin responds to ultraviolet rays at 10^{16} Hz.; retinal cells to light at 10^{15} Hz.; most organisms pick up thermal radiation at 10^4 Hz. The periods of neural signals are between three and ten seconds. Probably all living things, down to the genes, show circadian rhythms. Lunar periods are widely spread among many species, as are also circannual rhythms. Some bamboos flower every seven or eight years.

The morphology of oscillators across this band of twenty-four orders of magnitude must necessarily vary substantially. The gradual broadening of the spectrum of biological rhythms has thus forced upon evolving life a division of labor. Such a division is evident in each organism and collectively among members of a species. Even more broadly, the multiplicity of the species is itself a division of labor in the enterprise of carrying on life.

The upper limit of biological-clock frequency was probably reached in the ultraviolet clock. The lower limit is reached in those physiological clocks whose rates of change asymptotically approach linearity. These changes are manifest in the aging process, a hallmark of advanced forms of organisms.

At just what level aging and growth enter the cyclic order of life is not easy to say but the transition seems to relate to complexity. Simple molecules certainly do not age; individually they are prototemporal, in aggregates they are eotemporal. Hydrochloric acid is never young, neither is it ever old. On first approach the same appears to hold true for a DNA structure; after all, it is but a molecule. But biomolecules are dimensionally immense. The chromosomal complex of a small virus, the bacteriophage T4 (a virus that parasitizes bacteria) has some 200,000 nucleotide pairs. The chromosomes of aquatic animals have between 10^{11} and 10^{12} nucleotide pairs, those of man and all mammals about 5×10^9 pairs. Biological molecules can reach the molecular weight of 10^8 or perhaps even 10^{11}. Unlike small, inorganic molecules, these vast structures are never closed. Because of their need for maintaining internal coordination they may be regarded as the highly developed descendants of the first modest clockshop that determined biotemporal conditions.

It is to the biomolecular level that organisms return once in every life cycle. In reproduction, development, and aging we recognize a recapitulation of time's rites of passage. Every human being once in his or her lifetime determined a biotemporal umwelt, not very far above the pure succession of the physical world. Organisms that replicate do not go through a similar life cycle of descent and ascent.

Let us grant that directed time could now evolve. The physiological present of a living organism could now serve as a reference with respect to which future and past may acquire meaning. But the rhythm of life on earth is not one of an individual. Right now it is 6:40 A.M. not only for bakers but also for bears and buzzards. How can all things on earth, including man, act according to a common present unless that present is determined by a universal passage of absolute time in which all objects equally partake?

Just as individual organisms define their living present by maintaining their inner coordination, so do groups of living organisms: molds, mice, and men. If a coordinated, collective action is not possible, a common, collective present is not manifest; it does not exist, it has no meaning. Information exchange for the purpose of creating the communal present may travel by physical and chemical means but it must travel. Its rate of propagation determines the degree of coherence that a group of organisms can maintain. For instance, an earthwide present of 10 P.M. has no meaning unless by some sufficiently rapid means, such as radio signals, all the individual presents can be coordinated. On a more modest scale ants do it through chemicals, birds, mainly by voice.

Biosemioticians think of organisms as autonomous vehicles of signs and signals that, upon certain kinds of excitation and after a lapse of time, will take some action. For macromolecular forms of life the delay may be a few milliseconds or less. For simple metazoa it is minutes or hours. For individual animals it may be hours or months. Genetic responses may take millennia. Such delayed responses amount to a behavior which we usually ascribe to final causation or goal-directedness.

During autumn the daylight hours shorten. Certain birds along James Bay begin to display migratory restlessness and take off for warmer climes. The delayed action taken by the organism is the settling of the bird in a new nest along the northern shores of South America. Interpreted instant by instant in the living present, the behavior appears purposeful even if the bird does not see the new nest in its mind's eyes. It makes its decisions in terms of the historical sediment of contingencies, known as instincts. The fact remains that its actions are goal-directed, unlike the eruption of a volcano. Final causation, as asserted earlier, is the connectedness appropriate among events of the biotemporal world.

7.2 Complexity

The idea of complexity has come up in this monograph again and again. The meaning of the term was assumed to have been intuitively obvious; it was not otherwise defined. On his work on automatons, discussed in the preceding section, John von Neumann proposed to measure complexity by the "crudest possible standards, the number of elementary parts." Let us refine this specification by adding the demand for stability as a way of separating the wheat of stable objects from the chaff of metastable ones, and examine the following proposal.

Let the complexity of an integrative level be defined as, and be measured by the number of stable, distinct structures that may be identified as belonging in that organizational level. We are not concerned with what nature could have constructed but only with what actually did evolve.

The number of stable atemporal objects is three: photons, gravitons, and neutrinos. If neutrinos turn out to have mass and must travel therefore, at subluminal speeds, the number of atemporal objects will be two. If the reader wishes to distinguish among electron neutrino, muon neutrino and tau neutrino, the number may change to two or five.

Among stable prototemporal objects we class all those particles of nonzero restmass that live longer than the atomic chronon. The number of such objects depends on what is selected as distinct. A figure of slightly over 200 may be justified if one includes all particles and all antiparticles as distinct, as well as particles of the same species but with different charges and strangeness.

The stable units of the eotemporal world, of the macroscopic matter of the universe are the chemical compounds. Chemists recognize some 1,500,000 stable compounds. A diligent group of theoreticians may be able to double or triple this figure, so let us assume that there exist between 10^6 and 10^7 stable chemical compounds. The complexities of the different levels, if measured as suggested, will be seen to be so profoundly different that orders of magnitudes are all with which we need be concerned.

The number of stable structures of the biotemporal world would be the number of distinct individual organisms that have lived thus far. Among sexually reproducing species the counting of each individual as a different stable structure is well warranted. Denbigh has stressed, in connection with his exploration of the ubiquitous inventiveness of nature that, because the number of combinations of heterozygous genes vastly exceeds the number of members of a species, each member of the species, apart from identical twins, is very likely to have a distinct genotype.[19] The issue of distinctness gets more problematic as we approach the simplest forms of life, but identity, such as exists among elementary particles, cannot be found among the living. Making an estimate remains difficult, but for reasonable assumptions and leaving a rather phenomenal margin for error[20] one obtains a range between 10^{30} and 10^{40}.

Because of the large numbers involved, it is useful to change to the logarithms of the number of stable structures of an integrative level as measures of their complexities. Table 2 gives the summary of the figures.

I believe that the qualitative differences among the temporalities of the stable integrative levels of nature derive from the radically different complexities of those levels, as represented by the figures of Table 2.

7.3 The Usefulness of Number

The authority of physics rests on the power of quantified knowledge. While discussing time in the physical world this authority has not been questioned. On the contrary, with constant appeal to the generalized umwelt principle, it has been stressed again and again that number and abstractions are the only means through which we can hope to understand the behavior of light, particle, and galaxy. But as we begin considering the life process and work our way toward problems of noetic time, the issue of possible limits to the usefulness of quantitative reasoning must be raised. The answer is not at all self-evident; it must be developed.

What are the sources of the "unreasonable effectiveness" of number in natural science, to use Eugene Wigner's concise phrase?[21]

As the infant grows it constructs models of the world. In each model, events are connected causally, but the nature of causation changes. Judging from early behavior, the infant's first notion of causation, his first cosmology, is a belief in the efficacy of magic gestures.[22] This is a form of statistical, prototemporal connectedness wherein certain actions tend to bring forth certain results, on the average. The next level of connectedness is deterministic, mediated by the kinesthetic powers of the infant. The hitting of a small bell always gives the same music.

The operational space of the infant gets filled with objects

Table 2. Rough measures of the complexities of the stable integrative levels of nature

Integrative level	Number of stable structures (N)	Log N	Complexity (rounded out)
Atemporal	2–3	0.47	0
Prototemporal	200	2.3	2
Eotemporal	10^6–10^7	6–7	6–7
Biotemporal	10^{30}–10^{40}	30–40	30–40
Nootemporal*	10^{10^9}	10^9	10^9

* Entered here but derived in Section 7.4.

maintaining their identities for increasingly longer periods of time. Eventually the self of the child himself becomes an enduring identity. Leaving behind his infantile monism he identifies himself as his "one's self" and learns to live with the idea of one and the many. There are good reasons to believe that the continuous, unbroken symbol of the self is the source of the idea of oneness. It is interesting that for Pythagoras and for everyone else through the sixteenth century, "one" was the root of every number but not itself a number. Two was the least member of the multitudes of number.[23]

To know the world in terms of distinguishable, enduring structures is a necessary step toward being able to count, and counting is necessary before experience can be quantified. Similar skills make it possible to abstract invariances from feelings and from other sensations whose external sources cannot be identified and must be regarded, therefore, as nonspatial. The interweaving of the enduring elements of feelings with the description of objects gives human language its great power. From its rich matrix people have learned to separate what they judge to be necessary from what they deem as contingent features of reality. Being able to make this distinction is a necessary condition for the writing of quantifiable laws of nature.

According to developmental psychologists, first we learn to compare, then to evaluate. The child must first learn how to group objects by their lasting identities: all apples as distinct from all children. The notion of grouping is acquired as he learns matching one-to-one by touch counting. Next the child learns that the spatial rearrangement of a set of objects does not alter the sign, *number,* attached to the group. Finally, numbers as classes of abstract objects are discovered as one learns to deal with the principle of conservation of number: not of one particular number, but all numbers. The group now acquires the formal name *set,* with the single objects retaining their identities as its elements.

When a set is compared with other sets, matched one to one and found equivalent, the two sets are said to have the same cardinality. Two or more sets of the same cardinality may have different elements but each set in itself can contain only such elements as are, for the purpose of the cardinality of the set, indistinguishable. We already know that indistinguishable objects may be counted but cannot be placed in sequential order. Neither is this necessary for the determination of the cardinality of a set. The suggestion comes to mind that cardinal numbers somehow represent the temporal character of a world whose objects may be counted but not arranged by number: the prototemporal world.

Ordinality is discovered when the child is able to assimilate the relationship "more than" and "less than." What he learns to handle and to live with are relationships of two-way symmetry. Natural numbers arranged in an increasing sequence are also arranged thereby in a decreasing sequence. There is nothing in that ordering to make one direction preferable to the other. The order type of a well-ordered set is called an ordinal number. A well-ordered set of given ordinality must contain as many distinguishable members as is its ordinality, because each of its members must be uniquely designated by place, as if by sandwich boards hung on its shoulders. Because of the hierarchical relationship between ordinality and cardinality, an ordinal number designates both the order of its elements and the cardinality of the set. The suggestion comes to mind that ordinal numbers somehow represent the temporal character of a world whose events may be counted as well as arranged by number, though showing no preferred directions: the eotemporal world.

Zero as a place value notation came into use during the Seleucid Dynasty of kings in the Near East, two or three centuries before Christ. But the use of zero as a reference point about which two semi-infinite rays extend, was not possible until it was combined with the discovery of negative numbers. These first appeared in thirteenth-century China, then in sixteenth-century Europe. In the development of the mathematical ability of the child the use of zero and negative numbers also begins only after the mastery in counting and ordering has been achieved.

The real numbers (zero, positive, negative) suggest themselves representing the temporal character of the living world, with zero standing for the present. Assigning negative numbers to future events and positive numbers to past events, or vice versa, may certainly be useful for bookkeeping, but there is nothing in the behavior of ordered real numbers to correspond to final causation and purpose. A mathematical representation of connectedness by intent and need is yet to be developed. It may come, perhaps, from the study of certain mathematical operations easily performed in one direction but almost impossible to perform in the other direction, used in cryptography. These cryptographic methods employ the combination of very large numbers, suggesting that the one-directedness of the mathematical operations may relate to issues of complexity.

In any case, there is a need here for a break in the way mathematics is used. The radical nature of the necessary break is suggested by the difference in complexity between the eotemporal and the biotemporal integrative levels, represented by the numbers in Table 2.

There certainly seems to exist an isomorphism between the behavior of number on the one hand and, on the other hand, our understanding of the level-specific temporalities. It is a fair conjecture that the reasons for the isomorphism between number and experience reside in the psychobiological functions of the human brain itself. Unlike the evolution of biological structures in which earlier structures are altered, replaced, or embodied in later forms, the functions of the brain—by popular shorthand notation, the mind—retains the history of its own development. It is possible to identify among mental functions distinct organizational principles corresponding to the different temporalities.[24]

Mental work in the form of signs and signals, such as the symbols of mathematics, amounts to reports about the organizational affairs of the brain. Because the brain is made of inorganic matter, reports about the lower organizational principles of the brain are statements about the physical world which the brain, as a physical structure, shares with all other physical objects. If this be admitted as likely, then it will not be surprising to learn that the difficulties of reporting about time begin to grow substantially as those reports become increasingly self-referential. What is being approached here is the self-description of self-description, which is the stuff of art at its highest sophistication.

In spite of the jump of thirty orders of magnitude in complexity from the physical to the biological integrative level, mathematical description of life phenomena is not a hopeless enterprise, in principle, even though thus far the life sciences have just begun to try. I shall argue, however, that along the journey, the character of scientific and mathematical inquiry itself is likely to undergo a thorough and radical change.

Since the time of Archimedes, Euclid, and even Newton, the variety of notions and objects of interest to mathematics have increased tremendously. But the mathematical method remained the same. It is comprised of the observation of mathematical objects and of the abstraction from these observations the unchanging properties of the objects, so judged by the sense of time of the mathematician. The method of abstraction itself is not of interest to mathematics, neither is the machinery of contemplation, producing from the abstractions a small number of axioms. Well-defined sets of rules are then applied to the axioms so as to arrive at theorems and, by combining theorems, to obtain theories. Having followed religiously this method of thinking, mathematicians have come to be convinced that their findings, to be validated as truths, are subject only to the test of logical consistency. Logic itself was held to reflect a set of unchanging and inviolable rules, just as were the tenets of Euclidean geometry. On these solid foundations stands, or rather, stood, the structure of mathematical science.

In 1931 Kurt Gödel demonstrated that certain inherent limitations of the axiomatic method rule out the full axiomatization even of the ordinary arithmetic of integers, let alone the axiomatization of more complex systems.[25] The mathematical principles of this work are known as Gödel's incompleteness theorem. With its surprising direction it ruined the security offered by the logistic thesis. It revealed, instead, that mathematics is an open-ended hierarchy of inductively coupled deductive systems.

In his work on automata, mentioned in section 7.1, John von Neumann maintained that his findings on the complexity threshold necessary for successful self-reproduction were in harmony with Gödel's incompleteness theorem as well as with Russell's theory of types. We must imagine each level of complexity as associated with a different language, possessing a level-specific logic and a limitation on its capacity to define truth. Truth, in this context, is an expression of an unchanging law of nature. In the words of von Neumann:

> In the complicated parts of formal logic it is always one order of magnitude harder to tell what an object can do than to produce the object. The domain of validity of a question is of a higher type than the question itself. . . .
> . . . The feature is just this, that you can perform within the logical type that is involved everything that is feasible, but the question of whether something is feasible in a type belongs to a higher logical type.[26]

In Kurt Gödel's opinion what von Neumann referred to was the fact that a complete epistemological description of a language *A* cannot be given in the same language *A,* because the concept of truth of the sentences of *A* cannot be defined in *A.* It is this theorem which is the true reason for the existence of undecidable propositions in the formal systems containing arithmetic.

Gödel goes on to restate the epistemic issues involved in the hierarchy of complexities by stressing that "the description of what a mechanism is doing in certain cases is more involved than the description of the mechanism, in the sense that it requires new and more abstract primitive terms, namely, higher types."[27]

Gödel's theorem does not tell us that we will never be able to convince ourselves of the truth of one proposition or another, but only that there must be some whose validity cannot be proven and hence remains undecidable.[28] A decision may then still be reached but only by using a higher-order language which, by the earlier argument, will have its own limits of decidability.

Gödel's message about the hierarchical organization of mathematics along distinct levels of order and von Neumann's observation on self-reproducing automata both speak about hierarchical relationships among the laws, languages, and connectedness of different levels of complexity. Upon each level there are phenomena which, though their functions are appropriate to the level, cannot be recognized as systematic, lawful behavior except in terms of a higher-order language.

Consider now Table 1 in section 2.2. Special relativity theory, primarily the law of the atemporal world, is essential for the writing of quantum theory; yet it is impossible to predict the quantum properties of matter from the principle of relativity and the corresponding formalism. The atemporal world leaves the specifications of the prototemporal world undetermined. Likewise, quantum theory leaves undetermined the behavior of massive, gravitating matter, even though general relativity theory could not conclude anything contrary to quantum laws. The languages of the physical world leave undetermined the peculiar functions of the life process; the language of the life process leaves undetermined the forms of mental processes of which the human brain is capable.

Upon each organizational level there are more elements with which nature can play; each level is increasingly complex. This is probably the reason why the regions left undetermined upon each level by the lower-order languages contain increasing degrees of freedom.

I submit that mathematics appropriate to the handling of the level-specific issues of the higher integrative levels will inevitably loose its cherished rigor. The methodologies of its upper reaches may become indistinguishable from the kind of qualitative reasoning that students of life, mind, and society have already found necessary for their métier.

Consider, for instance, that branch of mathematics which analyzes problems of conflicts through the study of theoretical models of strategic methods; it is known as the theory of games. A classical theory of probability would limit the treatment of conflicts to probabilities. Those working with game theory maintain that their games of strategy open the way to mathematical attacks upon problems in economics, psychology, sociology, politics, and war by allowing for incomplete information, conflicting interests, and the interplay of rational decision and chance. Certain assumptions must, of course, be made about the behavioral patterns of the participants and about their interests.

Could the strategy of games be applied to the most inclusive game of all, the emergence of successive integrative levels?

We could not describe completely enough the behavior of the structures on level A unless we knew its laws, or most of its laws. But we have learned that a complete epistemological description of the laws of A (even if we could know most of them) cannot itself be given in the language of A. To obtain a concept of truth applicable to A we must appeal to B for the next higher level language, or family of laws. Thus, before we can make sufficiently exhaustive and reasonable assumptions about the behavior of the elements of level A, we must already know the language of B. Without knowing language B, we cannot even judge the domain of validity of the laws of A, let alone the domain of validity of the laws of B.

This is what is meant by the emergence of the unpredictably new. What assumptions can be made about the behavior of elementary particles to be able to predict the DNA molecule? About the DNA molecule to predict man? About man to predict his future history?[29] Whatever assumptions, and there are some, that can be made in these respective domains, they are likely to concern the least telling aspects of the unfolding of life, mind, and society.

Two decades ago, J. R. Lucas of Oxford University reflected upon the functions of an idealized computing device, known as the Turing machine, and on the teachings of Gödel as they apply to complex machines.[30] He concluded that above a certain level of complexity all machines will be increasingly unpredictable.

Mathematics appropriate for the operational analysis of these machines will have to allow, therefore, for various degrees of unpredictability. The engineer building computing devices of ever-increasing complexity will resemble Eddington's version of Robinson Crusoe. We know from Daniel Defoe that one morning Robinson Crusoe found a footstep in the sand. Then, Eddington tells us, he spent years of painstaking work in reconstructing the likely features of the visitor. Finally, when he beheld in his mind the image of the stranger, he recognized himself.

7.4 From Simultaneities of Need
to Expectations and Memories

During the evolution of the cyclic order of life some of the internally generated cycles, to which no prior, external cycles needed to correspond, were likely to have become manifest as behavioral cycles. Thus, the success of organic evolution gradually enriched the cyclic spectrum of rhythms to which the various species had to adapt. The increasing complexity of living clockshops combined with delayed reactions to make the behavior of organisms progressively less predictable. Fixed behavioral controls became insufficient because they had to meet increasingly unpredictable challenges—at least for some species in certain regions on earth.

Sometime after the human brain developed and before the appearance of the first human groups—so I conjecture—one or more forms of the genus Homo learned to work out future behavioral strategies based on past individual experience. The means that evolved and made such symbolic manipulation of experience possible is human language. This step brings to mind once again von Neumann's complexification threshold. Above but not below a certain level of the complexity of his brain an individual of the species H. sapiens was able to assist conspecific individuals to become better adapted than himself. Through language it became possible to create a store of collectively generated cumulative knowledge, comprised of acquired characteristics. From the biotemporal umwelt emerged the nootemporal world with its peculiar dynamics.

Through intricate evolutionary steps whose details are not known but whose outlines one may suspect, the development of the human sense of time seems to have followed along stages identical with the evolutionary stages of time itself. For the purposes of this monograph it is sufficient to note certain striking similarities between the organization of time perception in man and the hierarchy of temporalities in the universe.

We refer again to the structured Minkowski diagram of section 3.2 and inspect the area near its origin. From the point of view of physics, approaching the origins amounts to descending from the eo- to the prototemporal world, hence to the region where the distinction between spacelike and timelike geodesics becomes ill defined, and then entering atemporal chaos. An identical order of the different temporalities may be identified in the organization of psychological time, though sans geodesics.

Figure 13 shows the topology of the microstructure of human-time perception, based partly on the work of Ernst Pöppel.[31]

Stimuli separated by less than about 2 milliseconds will appear to human subjects as subjectively simultaneous. Whatever happens within this short interval can have no temporal structuring. This period, below the psychological fusion threshold, is the perceptual chronon. Two bursts of sound or other homogeneous or heterogeneous stimuli, separated by more than about 2 milliseconds usually enable the human subject to perceive them as distinct. But they must be separated by at least about 20 msecs. before their order can be correctly and reliably recognized. Between the fusion threshold as a lower boundary and the order threshold as its upper boundary the perceptual umwelt is prototemporal.

Above a certain period of time—20–50 milliseconds, depending on perceptual modality and mood—the topology of time perception becomes very intricate.[32] It is possible, however, to identify a period of time longer than the order threshold, though still too short for a conscious response, such that actions taken during this period remain reflexive and unconscious. Whether the corresponding umwelt is said to be eotemporal or biotemporal depends on the taste and judgment of the experimenter. Finally, there will be a certain minimal length of time necessary for a subject to take conscious action in response to a stimulus, such as to make a choice according to learned values.

Thus, the temporal umwelts in the structure of human-time perception form a hierarchy of nested presents, corresponding to the temporalities postulated by the principle of temporal levels. They are all subsumed in the mental present.

The perceptual chronon is contiguous with the atemporal boundaries of the physical world in the following sense. Beginning

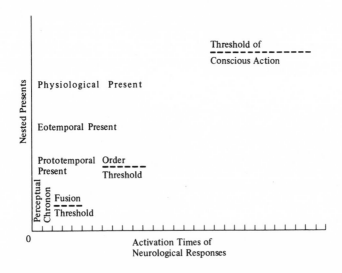

Figure 13. Topology of the microstructure of human-time perception. The hierarchical ordering of temporalities is identical with that of the structured Minkowski Diagram.

with that immensely complex system known as man, one can identify, or at least imagine, systems of lesser and lesser complexity, each with its atemporal limit, until one arrives at the natural limit of the atomic chronon. An analogous continuity exists for the proto- and eotemporal components of the mental present. However, nothing in the temporal structure of the inorganic world is contiguous with the physiological and mental presents. The physiological present may be made contiguous with those of other living organisms, the mental present with those of other humans.

The fundamental distinction between the noetic and biotemporal umwelts has nothing to do with the different accuracies with which physiological clocks and mental estimates can measure time intervals. The difference is qualitative. It resides in the peculiarly human capacity of imagining possible and impossible futures, factual and counterfactual pasts, and possible and impossible processes and things. These are the actors playing on the stage of long-term expectation and memory. What characteristics of the human brain may be responsible for these gifts? We may take a hint from the writings of W. Ross Ashby.

> The question of "complexity" must play a dominating role in our attempts to understand the brain (whether natural or artificial), for once we leave the mechanisms we knew before 1940, we arrive at forms whose complexity increases with overwhelming rapidity. Most properties in them increase not as the volume or the mass but with combinatorial speed, so that the order of their increase is either e^n or $n!$ or much faster still.[33]

The minimal set of essential responses to token stimuli, found in the most primitive instinct-reflex mechanisms of living organisms, represent the action of a central nerve cord of hundreds of thousands of neurons. An organism with a fully elaborated central nervous system and brain, containing 10^5–10^8 neurons, can learn to handle particularities of its environment, such as remembering places. In organisms with 10^9–10^{10} neurons, behavior ceases to be rigidly programmed and socialization is a prolonged process. The key social fact of this level was described by E. O. Wilson as "a perception of history."[34] I would prefer to say that animals, able to do noncyclic planning, have some limited foresight and memory. The perception of history I would reserve for members of our species. But man is in the same league with the chimpanzee and the baboon when it comes to the number of neurons he has. Therefore, the difference in mental capacity between man and the higher apes cannot be accounted for by the number of neurons alone, nor simply by the difference of brain size.

From Lumsden and Wilson we learn something else to which the difference could not possibly be attributed.[35] They have shown, using the methods of information theory, why it is impossible to program a human-sized brain and language structure from the available gene complement alone. To possess an inborn vocabulary of 10,000 words and speak in sentences of 10 words would require a genome of 10^{40} nucleotides or 10^{16} kg. of DNA. What the genes can and surely do provide is the development of a structure—yet to be educated. The responsibility is thus shifted to the sociocultural integrative level.

A cubic centimeter of water contains some 10^{22} molecules. Borrowing a term from the biologist Elsasser's "statistics and the concept of immensity," we may say that the number of microstates of such an assemblage is immense.[36] But this immensity has no structural or functional significance because none of the microstates could be substantially different from any other microstates or even from the microstates of a sample of 10^{32} or 10^{132} water molecules. A gram of water is not an autonomous system; it is not coordinated internally, its elements are not interconnected in various configurations, it has no distinct macrostates.

The human brain contains an estimated 10^{10} neurons, each of them connected to 10^4 or 10^5 synapses which, in their turn, may be connected to any and all of the other synapses. Stafford Beer noted some time ago that although each synapse is an analogue system internally, it functions as a binary system in so far as it either does or does not produce a pulse in the axon.[37] He concluded, therefore, that the number of inner configurations possible for a human-sized and human-type brain is of the order of $2^{10^{10}}$ or 10^{10^9}. Each of these theoretically possible brain states may be assumed to correspond to a global engram, with each engram representing a stable map of the world at an instant.

The limits of human memory for a lifetime of nothing-but-learning has been estimated to be 10^9 or 10^{10} bits. [38] But the power of the mind is not in its capacity for memory alone but in its ability to recognize and create correlations among chunks of memory. For phenomenal memories we build computers; for imaginative thought we turn to humans.

Because the genetic endowment is insufficient, by several orders of magnitude, to produce what is often called a hardwired brain— a despicable borrowing from the engineer—from where do the organizing principles come? Instead of expecting to produce a hardwired brain from genes, we must look at the noetic world and realize once again that its information content resides in language, art, and artifact. It is socialization and civilization which changes the ready brain of the infant to the mature brain of the adult. And if this be anything new, "I never writ, nor no man ever lov'd."

Let us consider the 10^{10^9} figure as the potential upper limit to the number of stable states that the human brain is capable of assuming, and so enter it in Table 2, section 7.2. If the brain were totally "wired," it would have $10^{10}(10^{10} - 1)$ or 10^{20} connections. Students of hierarchy theory maintain that a totally connected complex system is unstable. They define connectedness as the degree to which each element is connected with all other elements. Gardner and Ashby considered the "connectance" of large cybernatic systems and concluded that as the number of elements increases, there appears a sharp cutoff as a function of the degree of connectedness.[39] At a certain level of connectedness the system becomes precipitously unstable through self-generated catastrophe.

The isomorphism between the behavior of their model and the known behavior of the mind is so suggestive that it is worth mentioning even if it turns out to be spurious—and perhaps it is not. The ever-present possibility of self-generated catastrophe may well be the condition that the unity of conscious experience must control. The loss of that unity is the corollary of mental breakdown. The more creative a mind, the closer it usually is to the boundaries of collapse. It is rather a truism that the mind of the genius is often on this side of a thin wall whose other side is madness. We may rest assured that the brains of most people are far from being 100 percent "wired."

But the number of potentially distinct states of the human brain is still the same as for the baboon. How did members of our species learn to take advantage of the immense potentialities? What is the technique of the sociocultural process that promotes the "wiring" of the brain?

The human brain could have evolved into what it is now only if its capacity to generate and retain particular engrams was strongly selected for. It is unlikely that the selection mechanism involved any single behavioral traits or that it was linear; we have come all too far in all too short a time. It is more likely to have involved a syndrome of a number of coemergent functions, forming negative feedback and, by mutually reinforcing each other, hastening their collective development. Perhaps it involved the expansion of delayed action to long-term expectation; recall to long-term memory; the invention of language; and the discovery of the inevitability of death. Improved language skills increase the potential of memory storage; this makes possible an increase in the store of autogenic (internally generated) imagery, representing future contingencies. All these assist in an increasingly sharper definition of the self. Better delineation of identity in turn increases an awareness of death and thus promotes a sharper distinction between future and past.[40]

Let us review the limiting processes and structures of nature encountered in the preceding chapters. The immensely large, limiting object is the universe. The immensely small objects of physics are the particles. The immensely cold is $0°K$, the immensely hot is perhaps $10^{12}°K$. The limiting velocity is the speed of light. All these boundaries may be continuously approached from standard laboratory conditions, yet none of their peculiar properties is predictable from the laws that govern the world where a centimeter, a gram, and a second are practical measures. Without exception, the laws of these boundaries were all discovered much later than the laws of average sizes. The two sets, in each case, were tied together only ex post facto.

I believe that the unique qualities of the human brain, manifest in the behavior we ascribe to the mind, originate in the complexity of the brain. I also submit that the complexity of the human brain is a limiting condition of nature in the same class as the other limiting conditions mentioned are. It is neither surprising nor mystifying, therefore, if the laws governing mental functions are not obtainable from the laws of other biological structures by a linear, generalizing process. Instead, an appeal must be made to such nonbiological notions as selfhood, conscious experience, human freedom, and human creativity.

7.5 Distances and Times

The evolutionary development of time is accompanied by corresponding stages in the evolution of space.

It has been known, at least since the work of Hilbert, that mathematical analysis is competent to deal with spaces of any arbitrary dimensions. For instance, following Willard Gibbs, a 6-space may be used to record in a single continuous line the fate of a particle in terms of its three coordinate positions and three components of momentum. This kind of a multidimensional space is called phase, or parameter space. Two particles may be represented by a single line in a space of 12 dimensions, N particles in a $6N$ dimensional space. A single point in a space of 10^{90} can stand for the state of the universe at an instant, neglecting the problem of "the universe at an instant." A parameter space of a modest 100 dimensions can take care of the solar system including Haley's comet. The direction of time in each case must be obtained from a living organism and smuggled into the representation by appropriate instructions.

With the success of the four-dimensional representation of the eotemporal physical world, it has become tempting to employ spaces of many dimensions for improved bookkeeping of physical parameters. John A. Wheeler has gone to what may be the limit to the usefulness of multidimensional spaces. Superspace, as it has been called,

> is a manifold each of whose "points" is an abbreviation for one 3-geometry. A submanifold H of S is "the classical history of the geometry of space" when space has been started off under some particular set of dynamic conditions. It consists of all those spacelike 3-geometries that can be obtained as spacelike sections through one particular 4-geometry

which satisfies Einstein's classical field equations.[41] Superspace contains infinitely many points, with each point representing one, and only one 3-geometry. A collection of these points is superspace, with the directional character of time smuggled in from unspecified sources.

The usual presentation of the idea of superspace commences with the curious shapeshifting of space-time, recognized in the general relativistic relations among macroscopic matter, space-time, and motion. As geometrization is extended to the microworld, quantum mechanics requires that not one but an infinity of space-time frames be available, each with a different topology. As Davies had put it: "In the domain of the quantum theory the apparently concrete world of experience dissolves away among a melee of subatomic transmutations. Chaos lies at the heart of matter; random changes, restrained only by probabilistic laws, endow the fabric of the universe with a roulette-like quality."[42]

Superspace is the Platonic, geometrical container of this chaos. It is a formal structure for all potential conditions, the grand generalization of Riemann's program of one hundred and thirty years ago, for "constructing the concept of multiply extended magnitudes out of general notions of quantity" (sec. 6.2).

We place superspace in temporary storage and turn to the concept of genetic identity, discussed by Reichenbach but originating in the earlier work of the Gestalt psychologist Kurt Lewin.[43] Genetic identity or genidentity, signifies the distinctness of an object maintained through time, an unchanging identity, the permanence of Zeno's arrow with its head, shaft, and feather. In a world made of indistinguishable elements there can be no genidentities. Although I am entitled to think of a cloud of electrons as continuous in time, I am not entitled to think of individual electrons as having remained themselves. If one electron under an imaginary microscope were suddenly replaced by another, we could or would never know it. Quantum mechanics recognizes this fact of the prototemporal world and makes allowances for it. Genidentity can apply only to distinguishable objects, such as those determining the character of the macroscopic physical world. The eotemporal is the first integrative level capable of supporting enduring identities.

The demand that the structures of the eotemporal world maintain their identities is also a demand for the single-dimensionality of time. A universe of more than one temporal dimension would be a self-contradictory and hence nonviable world, for it could not guarantee the genidentity of objects. The selfsame thing, while remaining unchanged along one dimension could change into something else along another dimension. The single continuous time dimension that guarantees the possibility of objects with enduring identities is the cosmological common time of general relativity theory.

Combining the two lines of reasoning a new perspective comes into view about the position of superspace in the evolutionary development of spaces. Once again, let us consider each integrative level, in order.

In the atemporal world no meaning can be given to extension and hence to space. The world is pointlike.

The discontinuous space of the prototemporal world demands the kind of divine bookkeeping that becomes possible, in principle, through the use of superspace. But the name is misleading if one wishes to construct a rational hierarchy of spaces. There is nothing superior about superspace in the sense of having features more advanced than ordinary space does, along the evolutionary stages of space. On the contrary. Since it is the geometry of the primitive world of particles, of the prototemporal integrative level, it ought to have been called subspace.*

When crossing from quantum theory to macroscopic physics,

---

* Compare the judgment implicit in the naming of superspace with the naming of a certain artistic and literary style, surrealism. There is nothing *sur* about surrealism in the sense of being above realism. On the contrary, surrealism expresses the inarticulate modes of human feeling and thought out of which, through psychological and cultural filtering, the mature feeling and thought of realism arises. The fact that surrealism, superspace, and incoherent music are simultaneously popular in the paintings, physics, and musics of our epoch signifies a crisis of cultural distancing, perhaps to begin a new phase of human development. See the closing portion of section 7.6.

the correspondence principle demands that the two formalisms yield identical results. Likewise, the four-dimensional space-time of the eotemporal world emerges from the incoherent, statistical, and discontinuous aggregates of propagating probability waves, whose geometry is the sub-, alias superspace.

The three dimensions of ordinary space are probably as necessary for the maintenance of self-identical structures through time as is the single dimension of time, though compelling reasons for this argument are not easy to find. Nevertheless, consider that as a consequence of Huygens's principle pertaining to the wave propagation of light, an impulse propagating in three dimensions is quite different from an impulse propagating in two dimensions. In a two-dimensional world the impulse can have a sharp leading edge but its tail must trail, theoretically, to infinity. Flatland is a washed-out land electromagnetically; it is difficult to imagine how particles could form in it and gather into well-defined structures. Einstein's theory of gravitation would not work in flatland and hence we could not have a finite but unbounded universe. If the world were to contain well-defined structures whose behavior was to be governed by laws of relative position and motion only, then the number of spatial dimensions as it began to evolve from the infinite number of dimensions of superspace had to stop at three.

In the homogeneous, isotropic, and unbounded universe of eotemporal space nothing can correspond to the idea of *here,* just as in a world of pure succession nothing can correspond to the notion of *now.*

The space of subatomic particles is superspace. Out of it we may imagine emerging the less chaotic space appropriate for the elements and molecules. Continuous, macroscopic directions could acquire physical meaning only with the gathering of particles into solid structures, as the universe cooled. Among spatial structures solid crystals are the most ordered ones, providing well-defined relative directions and distances among their molecules. But a crystal is an iterative, cyclic structure without any need for central coordination in support of its integrity. Thus it could not be used for a definition of hereness.

In a process analogous to the slowly narrowing physiological present, from the first clockshops to the simplest organisms now alive, we postulate a slowly narrowing definition of *here* versus *there.* The most primitive organisms known possess spherical symmetries. The spatial umwelt of the 30,000 species making up the phylum of protozoa, with a membership far exceeding that of any other animal group, is isotropic. In the world of the radiolarida, a beautiful example of spherical symmetry, one can speak about distances from a center but not yet about preferred directions. The world's first X-axis was born when protozoa began growing tails, called flagella. And even flagella are not linelike. They are cylindrical spirals spinning about their axes.

Whatever the operational space of a species, a living organism can maintain its integrity and autonomy only if, within its system, interdependent events occur at precise relative locations and incompatible events are effectively separated in space. It is thus that the life process can define hereness and thereness in the homogeneous space of the general relativistic universe.

7.6 Mesoforms and Laws of Nature

The postulate of stable organizational levels implies as its corollary that mesoforms are unstable. Whether stable or not, evolutionary continuity demands the identification of mesoforms between each two adjacent integrative levels. This section intends to show that mesoforms may indeed be identified passim, but their metastability makes them rare entries in the inventory of nature. We conclude by assessing the epistemic significance of the rarity of mesoforms.

To find a mesoform between the atemporal and the prototemporal levels of the physical world one should seek a particle traveling with the speed of light but having finite restmass. Such a configuration of parameters is explicitly forbidden by special relativity theory. But in quantum theory there are no must-nots, only ought-nots. The atemporal boundaries of the physical universe are populated by objects violating, for very brief periods, a number of the well-established principles of macroscopic physics. It is among the virtual and short-lived objects in the world of subatomic particles where one may want to look for the missing mesoform. Thus far I know of no applicants for this position.

Mesoforms between the prototemporal and the eotemporal integrative levels would be massive objects, continuous in size from particles to watermelons to Himalayas, up to the size of an average galaxy, filling intergalactic space. But this is not the way the universe is. All the watermelons and Himalayas and moons have been gathered into the fundamental particles of the universe, the galaxies. And the gathering was done quite speedily. The universe, now 2×10^{10} years old, became matter dominated when it was no more than 10^4 years old, with galaxies beginning to form when it was around 10^9 years old, or about 5 percent of its present age.

To find a mesoform between the eotemporal and the biotemporal worlds one should demonstrate spontaneous generation: observe a living organism arise from nonliving matter, in the laboratory. Until the advent of microscopic research in the middle of the seventeenth century, spontaneous generation was held to be the normal mode of production for certain organisms. But abiogenesis, as it is also known, is quite incompatible with our understanding of the origins of life even though life is held to have come about from no life. The difference between abiogenesis and the modern idea of biogenesis is one of time scale.

There do exist species today that can alternate between crystal-like and organismlike forms, but they only take advantage of the best of both worlds, depending on environmental conditions. They are not examples of life coming about from no life, right in front of our eyes. Would DNA qualify as a mesoform? I do not think so. Bare DNA, the structure itself has no viability; viable DNA is always part of a cell.[44]

Transitional forms between the inorganic and organic worlds either must have evolved into stabler structures or else were devoured by our thoughtless ancestors, or both. Almost as a side effect, they are also likely to have changed the environment so substantially that another biogenesis became impossible. And the transition happened with amazing speed. Some believe that life may have arisen as rapidly as in 100,000 years, a mere 1/10,000th of the age of the earth at that epoch.

A mesoform between the biotemporal and nootemporal worlds would be an advanced organism which is neither ape nor man. But these, too, have vanished and probably for reasons similar to those that eliminated primordial life forms: either they rapidly evolved or they died out. But the type of temporal organization that may be speculatively associated with the developmental stage of life between ape and man survives in the unconscious functions of the mind. It separates the fully conscious processes of the mind from the biological processes of the brain, and thereby protects and helps maintain the distinctness of the noetic and biological functions in the unitary system of human behavior.[45]

Finally, one may seek mesoforms between the world of nation-states and cultures on the one hand and, on the other hand, the institutionalized semi-autonomous organization of a worldwide society. There are signs that our epoch already entered such a transient state.[46] If the past history of mesoforms is any guide, the period of the transition will be rapid.

When mesoforms are examined as a class they reveal certain common policies. One of these is their rapid passing. The dynamics of evolution seems to demand that they develop very rapidly. Another is that they separate two kinds of causations, or languages, or modes of lawfulness.

Consider now that a law of nature is just this: a formal description of rules which are believed to govern the behavior of matter—nonliving, living, thinking, and feeling. Laws of nature demonstrate the belief of their formulators in the existence of a dynamic order, one that conjoins in a systematic relationship the motion of the sun's shadow, the barking of the dog, and the coming of the friend. Because of the hierarchical organization of nature, the time, space, and connectedness of level-specific laws must necessarily change from level of complexity to different level of complexity. Therefore, what we recognize as scientific principles must themselves be emergent aspects of nature. But nomogenesis, the coming about of laws of nature, is to be sharply distinguished from a belief that the laws of nature may be changing.

What is actually observed is the existence upon each organizational level of a region unrecognized and hence undetermined by the laws of that level. Out of these regions arise, in the course of evolution, unpredictably new structures and processes governed by new, stable principles. The universal ordering principle, time, itself partakes in these rites of passage.

8

Unity and Multiplicity
in Physics

The grand unification of everything in physics, to use a phrase that has become popular, belongs to a metaphysical tradition known as *monism.* In its broadest sense it stands for the belief in the unity of reality, in one sense or in another. The classical protagonist of monism was Parmenides of Elea; a modern exponent was Spinoza. In contrast to the monists, pluralists maintain that there are many, perhaps infinitely many, ultimate substances and realities. Leibniz, William James, and Bertrand Russell are generally considered to be pluralists.

The hierarchical theory of time is both monistic and pluralistic. It is monistic because it admits the common material basis of everything there is although, being postrelativistic, it considers the concepts of matter and energy as interchangeable according to certain rules. The theory is also pluralistic. It sees evolutionary progress as having created a hierarchy of stable integrative levels, with the essential and unique features of each level unreducible to any of the features of the level or levels beneath it. According to this view, therefore, a conceptual and technical unification of physics will not be possible unless allowances are made for the hierarchy of temporalities, causations, and languages. But such a unification cannot be accomplished except from the perspective of *all* modern science, considered together.

Albert Einstein probably would have objected to the notion of a hierarchy of causations, though most physicists are ready to live with two kinds: one for quantum theory and one for everything else. It is much more difficult to admit to a hierarchy of temporalities. This chapter identifies the roots of the difficulties in the conservative way of thinking about time, within physical science. It will be argued that the received view of time, the belief that time is a single unstructured aspect of the universe, has been made untenable by the advances of scientific knowledge.

8.1 The Conservative Tradition in the Study of Time in Physics

Saint Augustine, writing at the end of the fourth century about the rules of number and the rules of wisdom remarked that, "although it is not clear to us whether number is a part of or separate from wisdom, or whether wisdom is a part of or separate from number, or whether they are the same, it is clear that both are true, and immutably true."[1]

Johannes Kepler, in his *Harmonice Mundi,* composed twelve centuries after the days of Saint Augustine, proclaimed that the Christian knows that the true and essential principles of mathematics are coeternal with God.[2] How did the Christian come to this understanding? What else is he supposed to know about mathematical principles simply by virtue of being a Christian?

Certain behavioral patterns demanded of good Christians clearly favored mathematics as the most rewarding way of knowledge.[3] The abstractness of mathematical truth and those of geometrical figures helps to get the mind off the desires of the body: numbers and geometrical shapes are less emotionally loaded than are human forms. Therefore, they are useful means in the service of a world view that encourages people to deal with their drives through sublimating them rather than satisfying them.

Protestantism, even more than Catholicism, has always preferred the most parsimonious expressions of feelings about truth and nature and without doubt, mathematics is the most parsimonious language there is. Skillful use of number in trade also brought the merchant increased prosperity, supporting the conviction of the faithful that his beliefs were correct. God had to be a mathematician, for Western mercantile economy as well as for Plato, though for different reasons. The systematic testing of mathematized hypotheses against natural phenomena amounted to a continuous reconfirmation of the timeless power of the Christian Creator, contrasting sharply with the transient power of the created.

Early in this century the influential German sociologist Max Weber remarked that the origins of natural science ought to be sought in the decided propensity of Protestant asceticism for empiricism, rationalized on a mathematical basis. "The favorite science of all Puritan, Baptist, or Pietist Christianity was thus physics, and next to it all those natural sciences which used similar methods."[4]

It was not only Christian practices but also certain theoretical predispositions of Christian theology through which the religious ambience of the West assisted the coming about, and subsequent progress of, the industrial and scientific revolutions. The theoretical stance I have in mind concerned a unique evaluation of man's affairs on earth, a view of the world known as salvation history.[5] The concept of salvation history arose from the preoccupation of the ancient Hebrews with a divine plan. Other religious and philosophical traditions perceived history in terms of recurring cycles. For the Hebrews history was one of linear progression, according to God's purpose. The dynamics of Hebrew existence was understood to reflect the tensions of a symbiosis between Israel and Yahweh. It was a drama, developing from well-defined beginnings—the Creation of the world and the Covenant—and progressing toward a final reconciliation.

The notion of salvation history was extended and elaborated by the followers of Christ who saw in his appearance a confirmation of God's plan for man. The birth and death of the Savior served as a reference epoch to which time could be traced from Creation on, and from which it may be traced into the future, to the Last Judgment.

The idea of linear time was born in the minds of the ancient Hebrew and early Christian thinkers when intentional behavior, a concept distilled from the vicissitudes and hopes of human life, was projected upon the destiny of man and upon the destiny of the universe itself. The working out of salvation history was the very reality of existence to the early scientific geniuses of Christendom. The metaphysical beliefs implicit in that drama were espoused by the makers of the industrial revolution who transmuted them into a belief in a future controllable by man for his own benefit, without any divine assistance. Mathematical predictability was consistent with such teachings, as was the practice of sublimation. Step by step, through the alchemy of developing ideologies, mathematics has acquired the authority of a necessarily true and necessarily complete report about the structure of the world. During the rise of a world view allowing man to be his own master, the thunder of Jehovah and the promises of the suffering Christ were transmuted to an epithet of our epoch: "Progress is our most important product."

How deeply the linear view of time and the high regard for

quantification are functions of the cultural environment may be seen if we compare the homeostatic and self-contained world of China with that of the unbalanced and open West. In imperial China there was little call for concepts such as salvation history; neither was there a need to seek and secure the future of the individual or of the Empire in mathematical formulas. Security was already vested in the Emperor. Although, wrote Joseph Needham, time remained inescapably real for the Chinese mind, subjective conceptions of time and metaphysical idealism never occupied more than a subsidiary place in Chinese thinking.[6] But a collective belief in the subjective existence of time together with metaphysical idealism were the necessary, though perhaps not sufficient, ingredients of the Renaissance birth of science and of its Newtonian synthesis.

The intellectual and religious heritage of the West thus demanded that there be a subjectively existing time, embracing all creation, and that this time have a purpose, a direction, an arrow. Furthermore, if time was to be understood at all, physical science was the only critical mode of knowledge, having the proper authority to explain it. For people born and raised in the Western heritage all ideas in harmony with these metaphysical beliefs tend to sound self-evidently true.

All these beliefs are explicit throughout the literature of time in physics. Roger Penrose spoke for the overwhelming majority of physicists when, in an essay, "Singularities and time-asymmetry," he defined his task as seeking an answer "to one of the long-standing mysteries of physics: the origin of the arrow of time." Forty-five pages of careful arguments and 114 bibliographical references later, all of them taken entirely from physical science, he concludes:

> Some readers might feel let down by this. Rather than finding some subtle way that a universe based on time-symmetric laws might nevertheless exhibit gross time-asymmetry, I have merely asserted that certain laws are not in fact time-symmetric—and worse than this, that these asymmetric laws are not yet known![7]

Intellectual courage, implicit in these remarks, has been the necessary ingredient of creative science. But sometimes nature does not oblige with phenomena consistent with received metaphysics.

It is possible to go further back than Kepler or even Saint Augustine. We can identify an older Western tradition, one that has contributed greatly to the Renaissance birth and subsequent growth of Western science and to which both Saint Augustine and Kepler were heirs. I am thinking about Platonism in general and about the Platonic division between timeless, eternal forms and the temporal world of the senses, in particular. The division resembles the Christian distinction between the eternal domain of God's sensorium or mind, and the transient domain of man. The dichotomy of this philosophical-religious view has been built into the way most people think about time in physics.

In the words of Eugene Wigner: "The surprising discovery of Newton's age is just the clear separation of laws of nature on the one hand, and initial conditions on the other. The former are precise beyond anything reasonable; we know virtually nothing about the latter."[8] The concept *laws of nature* stands for the eternal forms of Plato, for the order given at the instant of Creation and embodied today in our mathematized knowledge of the world. In short, it stands for the timeless. The concept *initial conditions* stands for all things of which Plato would say generation is the cause: for becoming, for whatever we learn through our senses, through experiment and experience, in short, for the temporal.

But this is not the way the cake of the physical world is cut when physics is actually done: witness the detailed reasoning of this monograph. As a result, modern physics is schizophrenic when it comes to the issue of time. On the one hand, the metaphysical underpinning of physical science, necessary in its epoch but now outdated, is being carried along by linguistic inertia.[9] On the other hand, physical practice has entered a new world of discourse, one that becomes coherent only if it is understood in terms of an evolutionary order of temporal hierarchy. The notion that some things are temporal and some are timeless has come to the end of its usefulness.

8.2 The Evolutionary Synthesis of the Sciences

Special relativity, quantum theory, and general relativity theory may be unified only if it is recognized that the temporalities, the spaces, and the causations appropriate to the different integrative levels of nature possess certain irreducible features. By the reasoning of the prior chapters, this leaves physics without directed time. But since any model of the world worth its name must accommodate our undeniable experience of passing time, the unification of the major theorems of physics cannot acquire authority unless it is integrated with the sciences of life, man, and society. The principle of temporal levels offers a suitable framework for such an integration.[10]

The history of science demonstrates that succeeding theories tend to cover increasingly larger fields of experience and, by so doing, reveal coherence among findings that earlier did not seem to have related one to another in any fundamental way. This, I believe, is the case for the principle of temporal levels. It offers a new sense of coherence among the various branches of physics. It also ties physics to the life sciences and to the sciences of man without threatening the methodological independence of the different fields of knowledge. The theory seems to possess just the right amount of flexiblity to provide diversity in unity.

The price paid for the synthesis is a recognition of the evolutionary nature of time, space, and causation.

The term, removing the degeneracy, signifies the demonstration that an apparently homogeneous condition possesses a structure. It is a concept broadly used in particle physics though it has its origins in mathematics. The principle of temporal levels may be said to remove the degeneracy from the idea of time because it reveals that time, which used to be thought of as an apparently homogeneous, universal aspect of nature, is, in fact, a dynamic, developing, and open-ended hierarchy of temporalities.

The arguments in favor of this new view of reality are powerful. *Quod erat demonstrandum.*

Hickory Glen, Connecticut
September 10, 1981

Notes and References

1. The Measurement of Time

1 Plato *Timaeus* (trans. Benjamin Jowett) 37d–37e
2 Aristotle *Physics* (trans. W. D. Ross) Bk. 4–219b
3 Christiaan Huygens, *Horologium Oscillatorium* (Paris, 1673), Bks 1 and 5.
4 Henri Arzeliès, *Relativistic Kinematics* (Oxford: Pergamon Press, 1966), p. 21.
5 Albert Einstein, *Sidelines on Relativity* (London: Methuen, 1922), p. 36.
6 Herman Bondi, "Relativity," *Reports on Progress in Physics* 22 (1959): 105.
7 H. Salecker and E. P. Wigner, "Quantum limitations on the measurement of space-time distances," *Physical Review* 109 (1958): 571–77.
8 G. J. Whitrow, "An analysis of the evolution of the scientific method," *L'Age de la Science* 3 (1970): 279 (his italics).

2. The Principle of Temporal Levels

1 *Sir Isaac Newton's Mathematical Principles of Natural Philosophy and His System of the World,* trans. A. Motte, ed. F. Cajori (Berkeley: University of California Press, 1973), Bk. 3, "Rules of reasoning in philosophy," rule 4.
2 Jakob von Uexküll, *Umwelt und Innenwelt der Tiere* (Berlin: Springer Verlag, 1921), pp. 218–19. A good introduction to the *Umwelt Prinzip* is the translation of his *Streifzüge durch die Umwelten von Tieren und Menschen* (1934). It appeared under the title, "A stroll through the worlds of animals and men: a picture book of invisible worlds," in *Instinctive Behavior,* trans. and ed. Claire H. Schiller (New York: International Universities Press, 1957), pp. 5–80. Biological and psychological use has now naturalized the word *umwelt* into English.
3 For a sympathetic and enlightened introduction to semiotics and the work of von Uexküll, see T. A. Sebeok, *The Sign and Its Masters* (Austin: University of Texas Press, 1979).

4   J. A. Wheeler, "Beyond the black hole," in *Some Strangeness in the Proportion: a Centennial Symposium to Celebrate the Achievement of Albert Einstein,* ed. Harry Woolf (Reading, Mass.: Addison-Wesley, 1980), p. 352.

5   Horace B. English and Ava Champney English, *A Comprehensive Dictionary of Psychological and Psychoanalytic Terms* (New York: David McKay, 1964).

6   G. J. Whitrow, *The Natural Philosophy of Time,* 2d ed. (Oxford: Clarendon Press, 1980), p. 375.

7   "Hierarchy denotes that Occidental mode of thought which transmuted man's persistent inclination to assert order into a particular conception of the universe in terms of precisely-arranged levels of existence commonly termed 'degrees' " (*Dictionary of the History of Ideas,* ed. P. P. Wiener [New York: Charles Scribners, 1968–1973], s.v. "Hierarchy and Order").

8   *The Great Soviet Encyclopedia,* ed. A. M. Prokhov (New York: Macmillan, 1976), s.v. "Hierarchy," by L. A. Sedov.

9   J. H. Woodger, "The 'concept of organism' and the relation between embryology and genetics," *Quarterly Review of Biology* 5, no. 1 (1930): 1–22.

10 H. A. Simon, "The architecture of complexity," in *The Sciences of the Artificial,* ed. H. A. Simon (Cambridge, Mass.: The M.I.T. Press, 1969), pp. 86–118.

11 H. H. Pattee, "The physical basis and origin of hierarchical control," in *Hierarchy Theory: the Challenge of Complex Systems,* ed. H. H. Pattee (New York: Braziller, 1973), pp. 71–108.

12 Cyril Stanley Smith, "Structural hierarchy in inorganic systems," in *Hierarchical Structures,* ed. L. L. Whyte et al. (New York: American Elsevier, 1969), p. 83.

13 Joseph Needham, *Time the Refreshing River* (London: Allen & Unwin, 1944), "Integrative levels: a revaluation of the idea of progress," pp. 233–72. See also by the same author, "Hierarchical continuity of biological order," in his *Order and Life* (Cambridge, Mass.: The M.I.T. Press, 1968), pp. 109–69.

14 H. H. Pattee, "The problem of biological hierarchy," in *Towards a Theoretical Biology,* ed. C. H. Waddington (Chicago: Aldine, 1970), 3: 117.

15 V. Fock, *The Theory of Space, Time and Gravitation* (London: Pergamon Press, 1959), p. 375. For a summary of his views, see pp. 368ff.

16 S. W. Hawking and W. Israel, "Introductory survey," in *General Relativity,* ed. S. W. Hawking and W. Israel (Cambridge: Cambridge University Press, 1979), p. 21.

17 In spite of its immense significance, I will not deal here with questions of the sociotemporal umwelt or the concept of the collective present. The outlines of the temporality appropriate for a global community are only now coming into view. See J. T. Fraser, "Temporal levels in the social process: the integration of temporalities in social systems

modeling," in *Time, Cultures and Development,* ed. C. Mallmann and O. Nudler (New York: Pergamon Press, forthcoming).

18 A. S. Eddington, *The Nature of the Physical World* (Ann Arbor: University of Michigan Press, 1958), p. 99.

19 J. M. E. McTaggart, "The unreality of time," *Mind,* n.s. no. 68 (1908): 459 (italics added).

20 P. C. W. Davies, *The Physics of Time Asymmetry* (Berkeley: University of California Press, 1974), p. 3 (his italics).

21 Ibid., p. 27 (his italics). Also Olivier Costa de Beauregard, "No paradox in the theory of time anisotropy," in *The Study of Time I,* ed. J. T. Fraser et al. (New York: Springer Verlag, 1972), pp. 131-39.

3. Special Relativity Theory

1 The famous specifications for absolute time, space, and motion will be found in the Scholium of the Definitions, at the beginning of *Sir Isaac Newton's Mathematical Principles of Natural Philosophy and His System of the World,* trans. A. Motte, ed. F. Cajori (Berkeley: University of California Press, 1973).

2 Albert Einstein, "On the electrodynamics of moving bodies," in A. Einstein et al., *The Principle of Relativity,* trans. W. Perrett and G. B. Jeffrey (New York: Dover, n.d.), pp. 37, 38.

3 V. Fock, *The Theory of Space, Time and Gravitation* (London: Pergamon Press, 1959), see esp. p. 375.

4 A. S. Eddington, *The Mathematical Theory of Relativity* (Cambridge: At the University Press, 1960), p. 19.

5 Hans Reichenbach, *The Philosophy of Space and Time* (New York: Dover, 1958), p. 143.

6 Newton's beautiful argument is very seldom quoted at sufficient length to demonstrate its author's circumspect brilliance.
    "Absolute space, in its own nature, without relation to anything external, remains always similar and immovable. . . . Absolute motion is the translation of a body from one absolute space into another; and relative motion, the translation from one relative place into another. . . . But because the parts of space cannot be seen, or distinguished from one another by our senses, therefore in their stead we use sensible measures of them. . . . And so, instead of absolute spaces and motions, we use relative ones . . . but in philosophical disquisitions, we ought to abstract from our senses, and consider things in themselves, distinct from what are only sensible measures of them. For it may be that there is no body really at rest, to which the places and motions of others may be referred.
    "But we may distinguish rest and motion, absolute and relative, one from the other by their properties, causes and effects. It is a property of rest, that bodies really at rest do rest in respect to one another. And therefore it is possible, that in the remote regions of the fixed stars, or perhaps far beyond them, there may be some body absolutely at rest; but impossible to know from the position of bodies to

one another in our regions, whether any of these do keep the same
position to that remote body, it follows that absolute rest cannot be
determined from the position of bodies in our region" (Newton,
Mathematical Principles, pp. 7–9).

7 R. W. Brehme, "A geometric representation of Galilean and Lorentz
transformations," *American Journal of Physics* 30 (1962): 489–96.

8 Achieving speeds approaching *c* are not feasible by any presently
known technology. Imagining new methods of locomotion is a fine
activity but it can get uncomfortably close to science fiction. For
telling figures see L. Marder, *Time and the Space Traveler* (London:
George Allen and Unwin, 1971), pp. 108ff. The practical limitations
do not apply to slow transfers to distant stars involving, perhaps,
generations of humans. But in such trips special relativistic effects
would play only a minor role.

9 Eddington, *The Mathematical Theory of Relativity,* p. 23.

10 A beautiful, detailed treatment may be found in Henri Arzeliès,
Relativistic Kinematics (Oxford: Pergamon Press, 1966), pp. 154–64.

11 J. D. North, "The time coordinate in Einstein's restricted theory of
relativity," in *The Study of Time I,* ed. J. T. Fraser et al. (New
York: Springer Verlag, 1972), pp. 12–32.

12 R. V. Pound and G. A. Rebka, "Variation with temperature of the
energy of recoil-free gamma rays from solids," *Physical Review Letters* 4 (1960): 274–75. See also C. W. Sherwin, "Some recent experimental tests of the clock paradox," *Physical Review* 120 (1960):
17–21.

13 E. F. Taylor and J. A. Wheeler, *Spacetime Physics* (San Francisco:
W. H. Freeman, 1966), p. 160.

14 G. J. Whitrow, *The Natural Philosophy of Time,* 2d ed. (Oxford:
Clarendon Press, 1980), p. 10.

15 Karl Popper, "Irreversible processes in physical theory," *Nature* 181
(1958): 402–3.

16 W. Ritz, "Über die Grundlagen der Elektro-dynamik und die Theorie
der schwarzen Strahlung," *Physikalische Zeitschrift* 9 (1908):
903–7. Albert Einstein, "Zum Gegenwärtigen Stand des Strahlungsproblems," ibid., 10 (1909): 185–93. Their conclusions were summed
up in a brief communication by Einstein, using the same title as his
prior paper, ibid., pp. 323–24.

4. Quantum Theory

1 *The Collected Papers of Charles Sanders Peirce,* ed. C. C. Hartshorne
and Paul Weiss (Cambridge: Harvard University Press, 1931–35), 2,
par. 228.

2 Louis de Broglie, "Ondes et quanta," *Comptes Rendus des Academie
de Sciences* 177 (1923): 507–10; "Quanta de lumière, diffraction et
interferences," ibid., pp. 548–50. It is in the second communication
that de Broglie asserts, "La nouvelle dynamique du point matériel
libre est à l'ancienne dynamique (y compris cella d'Einstein) ce que

l'optique odulatoire est à l'optique geometrique" (p. 549). The symmetry is complete.

3 For instance, David Park, *Introduction to Quantum Theory*, 2d ed. (New York: McGraw Hill, 1974), pp. 47ff.

4 Niels Bohr, "The quantum postulate and recent development of atomic theory," *Nature* 121 (1928): 580–90. This is a revised version of a lecture delivered in 1927.

5 Ian Hacking, *The Emergence of Probability, a Philosophical Study of Early Ideas about Probability, Induction, and Statistical Inference* (Cambridge: Cambridge University Press, 1975) is as masterful as it is interesting.

6 E. J. Yarmchuk et al., "Observation of stationary vortex arrays in rotating superfluid helium," *Physical Review Letters* 43, no. 3 (1979): 214–17.

7 For experimental details see Claus Jönsson, "Electron diffraction at multiple slits," *American Journal of Physics* 42 (1974): 4–11. This is a translation by D. Brandt and S. Hirschi from the 1961 original. The appendix guides the reader to the post-1961 history of the experiment. For an early realization that the interference patterns may be independent of light intensity, see G. I. Taylor, "Interference fringes with feeble light," *Proceedings of the Cambridge Philosophical Society* 15 (1909): 114–15.

8 Park, *Introduction to Quantum Theory*, p. 57.

9 Jönsson, "Electron diffraction at multiple slits." Figure 7 is an electron diffraction pattern from a single slit, clearly showing the side bands.

10 A. Einstein, B. Podolsky, and N. Rosen, "Can quantum-mechanical description of physical reality be considered complete?," *Physical Review* 48 (1935): 777–80.

11 Niels Bohr, "Can quantum-mechanical description of physical reality be considered complete?," *Physical Review* 48 (1935): 700.

12 Ibid., p. 701.

13 For a summary discussion and as a guide to his earlier work, see J. A. Wheeler, "Delayed choice experiments and the Bohr-Einstein dialogue" (Papers read at the meeting of the American Physical Society and the Royal Society, June 5, 1980), pp. 11–40. The phrase quoted appears frequently in this and other writings of the author.

14 Ibid., p. 34.

15 For instance, P. E. Greenfield et al., "The double quasar 0957+561: examination of the gravitational lens hypothesis using the very large array," *Science* 208 (1980): 495–97.

16 R. P. Feynman, "The theory of the positrons," *Physical Review* 76 (1949): 749–59.

17 E. C. G. Stückelberg, "Remarque à propose de la création de paires de particules en théories de relativité," *Helvetia Physica Acta* 14 (1941): 588–94.

18 G. J. Whitrow, *The Natural Philosophy of Time*, 2d ed. (Oxford: Clarendon Press, 1980), p. 333.

19 David Park, *Classical Dynamics and Its Quantum Analogues* (New York: Springer Verlag, 1979).

20   P. C. W. Davies, *The Forces of Nature* (Cambridge: Cambridge University Press, 1979), p. 112.
21   P. C. W. Davies, *The Physics of Time Asymmetry* (Berkeley: University of California Press, 1974), p. 162.
22   Hugh Everett III, " 'Relative state' formulation of quantum mechanics," *Reviews of Modern Physics* 29 (1957): 459.
23   Ibid., p. 460.
24   Ibid.
25   Bryce S. DeWitt, "The Everett-Wheeler interpretation of quantum mechanics," in *Batelle Rencontres: Lectures on Mathematics and Physics* (New York: W. A. Benjamin, 1968), pp. 318–32.
26   E. P. Wigner, "Two kinds of reality," *Monist* 48 (1964): 251.

5. Thermodynamics

1   A. S. Eddington, *The Nature of the Physical World* (Ann Arbor: University of Michigan Press, 1958), p. 69.
2   Ibid., pp. 66, 79, 69.
3   The statistical meaning of the second law of thermodynamics as a trend in nature toward "mixed-upness" is beautifully illustrated in M. S. Watanabe's "Time and the probabilistic view of the world," in *The Voices of Time,* 2d ed., ed. J. T. Fraser (Amherst: University of Massachusetts Press, 1981), p. 529ff.
4   Sir Fred Hoyle took his base line on the nucleotide level. He calculated that the information content of higher forms of life is represented by the number $10^{40,000}$. This is the specificity with which some 2,000 genes, each of which might be chosen from 10^{20} nucleotide sequence of the appropriate length, might be defined ("Hoyle on evolution," *Nature* 294 [1981]: 105). He quoted the number in support of the well-known argument that organic evolution has not had sufficient time to produce the advanced forms of life. It had often been pointed out that large numbers so generated cannot be validly employed in reasoning against the historicity of evolution by natural selection. (For instance, H. Kalmus, "Organic evolution and time" in *The Voices of Time,* pp. 330–52). But they remain useful reminders of the immense complexity of living organisms.
5   See M. S. Watanabe, "Learning process and the inverse H-theorem," *I.R.E. Transactions, Information Theory* PGIT-18 (1962): 236–51; "A mathematical explication of inductive inference," *Records of the Colloquium on the Foundations of Mathematics* (Budapest: Akademia, 1965), pp. 67–107; *Knowing and Guessing: A Quantitative Study of Inference and Information* (New York: Wiley, 1969), chap. 5.
6   P. C. W. Davies, *The Physics of Time Asymmetry* (Berkeley: University of California Press, 1974), p. 54.
7   "Über den zweiten Hauptzatz der mechanische Warmetheorie," in *Max Plancks Physikalische Abhandlungen* (Braunschweig: Friedrich Vieweg and Sohn, 1958), pp. 1–61. He distinguishes between neutral and natural processes. A neutral process is one for whose final state

"die Natur die gleiche Vorliebe hat wie für den Anfangzustand. . . ."
A natural process is one for whose final state "die Natur mehr
Vorliebe hat wie für den Anfangzustand . . ." (p. 4). The same
emphasis on *Vorliebe* appears in "Über das Prinzip der Vermehrung
der Entropie" (1887), ibid., pp. 196–216. It again reappears in his *A
Survey of Physics* (New York: Dutton, n.d. [1925?]), p. 16. "Years
ago I took the liberty of describing it thus—that Nature has a greater
predeliction for the State *B* than for the State *A*." The idea of neutral
and natural processes comes up once again in his *Theory of Heat*
(London: Macmillan, 1932), p. 54. In modern terms, neutral processes
are called reversible; natural processes, irreversible.

8   For a precise definition of what is meant by dissipative structures and
how they are handled in nonequilibrium thermodynamics, see Ilyia
Prigogine, *From Being to Becoming* (San Francisco: W. H. Freeman,
1980).

9   For a concise and authoritative discussion, see K. G. Denbigh, *The
Thermodynamics of Steady State* (London: Methuen, 1951).

10   Joseph Needham, "Evolution and thermodynamics," a 1941 lecture
printed in his *Time the Refreshing River* (London: Allen and Unwin,
1943), p. 230.

6. General Relativity Theory

1   For a systematic comparison of Newtonian and general relativistic
cosmologies see G. F. R. Ellis, "Relativistic cosmology," in *General
Relativity and Cosmology* (New York: Academic Press, 1971), pp.
104–44.

2   *The Thirteen Books of Euclid's Elements,* ed. T. L. Heath (New
York: Dover, 1956), 1: 155.

3   The Schweikart-Gerling-Gauss exchange on astral geometry may be
found in Carl Friedrich Gauss, *Werke* (Leipzig: B. G. Teubner,
1900), 8: 177–82, and 10, pt. 2: 31–35.

4   Ibid., 8: 177 (my translation).

5   G. F. B. Riemann, "On the hypotheses which lie at the foundations of
geometry," *Habilitationsschrift.* In *A Source Book in Mathematics,*
ed. D. E. Smith, trans. H. S. White (New York: McGraw Hill,
1929), p. 411.

6   Ibid., p. 412.

7   A. Einstein, "On the influence of gravitation of the propagation of
light," in A. Einstein et al., *The Principle of Relativity,* trans. W.
Perrett and G. B. Jeffrey (New York: Dover, n.d.), p. 100.

8   This clever expository device is from J. J. Callahan's "The curvature
of space in a finite universe," *Scientific American* 235 (August 1976):
90–100.

9   P. T. Landsberg and R. K. Pathria, "Cosmological parameters for a
restricted class of closed big-bang universe," *Astrophysical Journal*
192 (1974): 577–79.

10   General relativity theory is held to be asymptotically valid down to
certain values of time, distance, density, and curvature. From com-

binations of the fundamental constants of physics so arranged as to yield the correct dimensions, one obtains for minimal values, $t = 5.3 \times 10^{-44}$ sec., and $r = 1.6 \times 10^{-33}$ cm. See, e.g., O. Godart and M. Heller, "La théorie du 'big bang' et l'hypothèse de l'atome primitif," *Revue des Questions Scientifiques* 147 (January 1976): 3–17.

11   Don C. Kelly, *Thermodynamics and Statistical Physics* (New York: Academic Press, 1973), pp. 399–404.

12   On the technical details see the masterful summary by Allan Sandage, "The Redshift," in *Galaxies and the Universe,* ed. Allan Sandage et al. (Chicago: University of Chicago Press, 1975), pp. 761–83.

13   David Layzer, "Galaxy clustering," in *Galaxies and the Universe,* p. 695.

14   J. R. Gott, "Computer simulation of the universe," *Comments on Astrophysics* 8, no. 3 (1979): 55–67.

15   G. J. Whitrow, *The Natural Philosophy of Time,* 2d ed. (Oxford: Clarendon Press, 1980), pp. 302–7.

16   Ibid., pp. 264–66.

17   If the universe does function as a dynamic unit undergoing secular change, that change might also be manifest in some ways other than the expansion. One may wonder, for instance, whether certain quantities ordinarily regarded as constant might not be varying with time? Examples might be the constant of gravitation, the charge on the electron or the ratio of gravitational to electric forces between the proton and the electron.
There is no evidence whatsoever that such changes do take place. Furthermore, the usual form of the question, "Are the fundamental constants of physics really constant?," hides an epistemic fallacy. If someone had demonstrated that the gravitational constant underwent a long-term secular change, a new expression would immediately be formulated: $G = G_0 \, (t)$, with G_0 taking over the role of the formerly constant G. The need to identify constant relationships in nature is built into the way physics is being carried on (see sec. 8.1).

18   Saint Augustine, *The City of God,* trans. and ed. Marcus Dods (New York: Hafner, 1948), pp. 441–43.

19   *The Confessions of Saint Augustine,* trans. E. B. Pussey (New York: The Modern Library, 1949), p. 365.

20   Saint Augustine, *The City of God,* p. 442.

21   For a summary, see M. G. Edmunds, "Open debate," *Nature* 288 (1980): 431–32.

22   Michael Heller, "The origins of time," *The Study of Time IV,* ed. J. T. Fraser et al. (New York: Springer Verlag, 1981), p. 91.

23   J. A. Wheeler distinguishes two kinds of time: a dynamic time (cosmological or quasi-Newtonian) and a statistical-biological time (valid for statistical aggregates and living organisms). He then turns to the question of time during the collapsing phase of the universe, and asks, "As dynamic time marches forward, what will happen then to the statistical and biological time? Will they continue to point in the same direction or will they point in opposite directions? In the

one case, to a person alive in the second phase of the universe, the universe will appear to be contracting. In the other case, it will appear to be expanding simply because a moving picture of contraction run backwards looks like expansion. Many colleagues agree that the question is open and that the answer is one of the great puzzles of our day. . . ." ("At last—a sane look at the 'arrow of time,' " *Physics Today,* June 1975, pp. 49–50).

24 P. C. W. Davies, *The Physics of Time Asymmetry,* p. 200.

25 R. D. Blandford and K. S. Thorne, "Black hole astrophysics," in *General Relativity,* ed. S. W. Hawking and W. Israel (Cambridge: Cambridge University Press, 1979), p. 502.

26 On these intriguing issues see J. D. Bekenstein, "Black holes and entropy," *Physical Review* D. 7, no. 8 (1973): 2333–46 and S. W. Hawking, "Black holes and thermodynamics," ibid., 13, no. 2 (1976): 191–97.

27 S. W. Hawking, ibid., p. 191.

28 N. D. Birrell and P. C. W. Davies, "On falling through a black hole into another universe," *Nature* 272 (1978): 35–37.

29 Peter Havas, "Four-dimensional formulation of Newtonian mechanics and their relation to the special and general theory of relativity," *Reviews of Modern Physics* 36 (1964): 938–65.

30 C. Callan, R. H. Dicke, and P. J. Peebles, "Cosmology and Newtonian mechanics," *American Journal of Physics* 33 (1965): 105–8. See also Peter Landsberg, "Q in cosmology," *Nature* 263 (1876): 217. For a recent, clear summary, see Wolfgang Rindler, *Essential Relativity* (New York: Springer Verlag, 1977), sec. 9.8 and related material, and Ellis, *General Relativity and Cosmology.*

31 P. C. Davies, *Other Worlds* (New York: Simon and Schuster, 1980), p. 198. For a full treatment see Davies's *The Physics of Time Asymmetry.*

32 David Layzer, "Cosmology and the arrow of time," in *Vistas in Astronomy,* ed. A. Beer (New York: Pergamon Press, 1972), pp. 279–87; "Galaxy clustering," in *Galaxies and the Universe,* pp. 665–723; "The arrow of time," *Astrophysical Journal* 206 (1976): 559–69.

33 Layzer, "Cosmology and the arrow of time," p. 286.

34 Ibid., p. 285.

35 Fraser, *The Voices of Time,* p. xvii.

7. Biogenesis and Organic Evolution

1 An up-to-date, coherent, and authoritative summary is Jürgen Aschoff, *Biological Rhythms* (New York: Plenum Press, 1981). For earlier work, see the first book-length study of physiological rhythms: Erwin Bünning, *The Physiological Clock,* rev. 3d ed. (New York: Springer Verlag, 1973).

2 B. C. Goodwin, *Analytical Physiology of Cells and Developing Organisms* (New York: Academic Press, 1976). Also by the same author, *Temporal Organization in Cells* (New York: Academic Press, 1963). J. T. Bonner, *On Development* (Cambridge: Harvard Univer-

sity Press, 1974). As an entry to the writings of Jürgen Aschoff, see chaps. 1, 6, 17 and 24 in *Biological Rhythms.* A. T. Winfree, *The Geometry of Biological Time* (New York: Springer Verlag, 1980).

3   *The Life and Letters of Charles Darwin,* ed. F. Darwin (London: Murray, 1887), editorial note, letter to Hooker, 29 March 1863.

4   Joseph Needham, "Integrative levels: a revaluation of the idea of progress," a 1937 lecture, printed in his *Time the Refreshing River* (London: Allen and Unwin, 1944), p. 255. Also, his *Order and Life* (Cambridge, Mass.: The M.I.T. Press, 1968), esp. p. 158.

5   J. D. Bernal, *The Origins of Life* (Cleveland: World, 1967), pp. 192ff; and J. D. Bernal, *The Physical Basis of Life* (London: Routledge and Kegan Paul, 1951), pp. 34ff.

6   G. H. Brown and J. J. Wolken, *Liquid Crystals and Biological Structures* (New York: Academic Press, 1979).

7   A. G. Cairns-Smith, "Beginnings of organic evolution," *The Study of Time IV,* ed. J. T. Fraser et al. (New York: Springer Verlag, 1981), pp. 15–33. The article also serves as an entry to the related literature.

8   C. E. Muses, "Aspects of some problems in biological and medical cybernetics," in *Progress in Biocybernetics,* ed. Norbert Wiener and J. B. S. Haldane (New York: Elsevier, 1965), pp. 243–48.

9   Ilya Prigogine, *Introduction to Thermodynamics of Irreversible Processes* (New York: Wiley, 1955), p. 92.

10 W. W. Forest and D. J. Walker, "Change in entropy during bacterial metabolism," *Nature* 201 (1954): 49–52.

11 Ilya Prigogine, *From Being to Becoming* (San Francisco: Freeman, 1980), p. 88. That the stationary state of an open-reaction system is not, in general, a state of minimum entropy production has been shown long ago by K. G. Denbigh, "Entropy creation in open reaction systems," *Transactions of the Faraday Society* 48 (1952): 389–93.

12 Manfred Eigen and P. Schuster, *The Hypercycle: A Principle of Natural Self-Organization* (New York: Springer Verlag, 1979).

13 M. C. Moore-Ede and F. M. Sulzman, "Internal temporal order," in *Biological Rhythms,* p. 215.

14 John von Neumann, *Theory of Self-Reproducing Automata,* ed. and completed A. W. Burns (Urbana: University of Illinois Press, 1969), pp. 79, 80.

15 For related details see J. T. Fraser, *Of Time, Passion and Knowledge* (New York: Braziller, 1975), pp. 178–231. See also by the same author, "Temporal levels: sociobiological aspects of a fundamental synthesis," *Journal of Social and Biological Structures* 2 (1978): 339–55.

16 P. T. Saunders and M. W. Ho, "On the increase of complexity in evolution," Part 1 in *Journal of Theoretical Biology* 63 (1976): 375–84; Part 2 in ibid., 90 (1981): 515–30.

17 For this little-explored region of molecular biology, see the writings of C. S. Pittendrigh. An entry to his work may be had through his paper. "Circadian clocks: What they are?," in *The Molecular Basis of*

Circadian Rhythms, ed. J. W. Hastings and H. G. Schweiger (Berlin: Abakon, 1976), pp. 11–48. See also the other papers in the same volume.

18 Manfred Eigen and Ruthild Winkler-Oswatitch, "Transfer-RNA, an early adaptor," *Naturwissenschaften* 68 (1981): 217–28 and "Transfer-RNA, an early gene?," ibid., pp. 282–92.

19 K. G. Denbigh, *Three Concepts of Time* (New York: Springer Verlag, 1981), p. 136.

20 It has been estimated that the number of insects per moist acre of forest floor is over a million. There are about 4 billion hectares of forest on earth at this time with the likelihood of a total of 10^{15} insects. Now, insects have been around since the carboniferous period of the Paleozoic era, that is, for some 260 million years. This leads to a total insect population, living or dead, of 10^{23}–10^{24} individuals. Let us guess that the number of organisms larger than insects that have ever lived are one animal or plant to every 1,000 insects. This figure hardly changes the 10^{23}–10^{24} range. On the other hand, there may be as many as 10^9 independent and distinct microorganisms of all kinds to be found inside and on the organisms counted. And there may be 10^6–10^7 bacteria per gram of soil. Put it all in a hat, shake it together, get a figure between 10^{30} and 10^{40}.

21 Eugene Wigner, *Symmetries and Reflections* (Bloomington: Indiana University Press, 1967), p. 222.

22 J. T. Fraser, "Temporal levels and reality testing," *International Journal of Psychoanalysis* 62 (1981): 3–26.

23 For a cultural history of number, relevant to this discusion, see Karl Menninger, *Number Words and Number Symbols* (Cambridge, Mass.: The M.I.T. Press, 1970).

24 J. T. Fraser, "Temporal levels and reality testing."

25 Kurt Gödel, *On Formally Undecidable Propositions of Principia Mathematica and Related Systems,* trans. B. Meltzer (Edinburgh: Oliver and Boyd, 1962).

26 von Neumann, *Theory of Self-Reproducing Automata,* pp. 47–48, 51.

27 Ibid., p. 55.

28 On the hierarchical nature of undecidability, see Martin Davies, "What is computation?," in *Mathematics Today,* ed. Lynn Arthur Steen (New York: Springer Verlag, 1978), pp. 241–67.

29 "If man could convince himself by natural science and mathematics that he is a predictable machine . . . he would once more, out of sheer ingratitude, attempt the perpetration of something perverse which would enable him to insist upon himself. And if he could not find means, he would contrive destruction and chaos, will contrive sufferings of all sorts, for the sole purpose, as before, of asserting his personality. He will launch a curse upon the world . . . to convince himself that he is not a machine. If you say that all this, too, can be calculated and tabulated—chaos and darkness and curses, so that the mere possibility of calculating it all beforehand would stop it all and reason would reassert itself—man would purposely become a lunatic, in order to become devoid of reason, and therefore able to insist upon

himself" (F. M. Dostoyevsky, *Notes from Underground,* trans. B. G. Guerney, in *Short Novels of the Masters,* ed. C. Neider [New York: Rinehart and Co., 1948], pp. 145–46).

30 J. R. Lucas, "Minds, machines and Gödel," *Philosophy* 36 (1961): 112–27.

31 Ernst Pöppel, "Time Perception," in *Handbook of Sensory Physiology,* ed. R. Held et al. (New York: Springer Verlag, 1978), 8: 713–29.

32 For a discussion of the issues involved, see John Michon, "The making of the present—a tutorial review," *Heymans Bulletins,* Rijksuniversiteit Groningen: Instituut voor Experimentele Psychologie, October 1976.

33 W. Ross Ashby, "The place of the brain in the natural world," *Currents in Modern Biology* 1 (1967): 103.

34 E. O. Wilson, *Sociobiology: The New Synthesis* (Cambridge, Mass.: The Belknap Press, 1975), p. 151.

35 C. J. Lumsden and E. O. Wilson, *Genes, Mind and Culture: The Coevolutionary Process* (Cambridge: Harvard University Press, 1981), pp. 333–41.

36 W. M. Elsasser, *Atom and Organism: A New Approach to Theoretical Biology* (Princeton: Princeton University Press, 1966), pp. 69ff.

37 Stafford Beer, *Decision and Control* (New York: Wiley, 1966), pp. 363–64.

38 J. S. Griffith, *Mathematical Neurobiology* (New York: Academic Press, 1971). See sec. 6.4.

39 M. R. Gardner and W. R. Ashby, "Connectance of large dynamic (cybernetic) systems: critical values for stability." *Nature* 228 (1970): 784–85.

40 Details of the early evolution of man's mental faculties are unknown. Considering that voiced symbols must have played a central role in that evolution, it is doubtful whether those details could ever be known by direct evidence. Instead, they must be reconstructed from whatever evidence there is. See J. T. Fraser, *Time as Conflict: A Scientific and Humanistic Study* (Boston: Birkhäuser Verlag, 1978), pp. 98–130. See also John C. Eccles, *The Human Mystery—the Gifford Lectures 1977–1978* (Berlin: Springer International, 1979), esp. lectures 5 and 6.

41 J. A. Wheeler, "Superspace and the nature of quantum geometrodynamics," in *Batelle Rencontres 1967: Lectures in Mathematics and Physics,* ed. S. DeWitt and J. A. Wheeler (New York: Benjamin, 1968), p. 247.

42 P. C. W. Davies, *Other Worlds* (New York: Simon and Schuster, 1980), p. 92.

43 Hans Reichenbach, *The Philosophy of Space and Time* (New York: Dover, 1958), p. 270.

44 J. T. Bonner, *On Development,* p. 75.

45 J. T. Fraser, "Temporal levels and reality testing."

46 J. T. Fraser, "Temporal levels in the social process: the integration of temporalities in social systems modeling," in *Time, Cultures and De-*

velopment, ed. C. Mallman and O. Nudler (New York: Pergamon Press, forthcoming).

8. Unity and Multiplicity in Physics

1 Saint Augustine, *On Free Choice of the Will,* trans. A. S. Benjamin and L. H. Hackstaff (Indianapolis: Bobbs-Merril Co., 1964), bk. 2, sec. 11: "How are the rules of number and wisdom related" (quote from verse 129).

2 Johannes Kepler, *Harmonice Mundi* 4:1. In *Johannes Kepler Gesammelte Werke,* ed. Max Casper (Munich: C. H. Beck'sche Verlagsbuchhandlung, 1940), 6: 219.

3 On this, see J. T. Fraser, *Of Time, Passion, and Knowledge* (New York: Braziller, 1975), pp. 321–60, and by the same author, *Time as Conflict* (Basel: Birkhäuser, 1978), pp. 204–37. The history of science and technology in its relation to intellectual history is being explored by Francis C. Haber in a book now in preparation.

4 Max Weber, *The Protestant Ethics and the Spirit of Capitalism,* trans. Talcot Parsons (New York: Scribner, 1904), 249 n. 145.

5 See the masterful works of S. G. F. Brandon. His "Time and the destiny of man," in *The Voices of Time,* ed. J. T. Fraser (Amherst: University of Massachusetts Press, 1981), pp. 14–57, may serve as an introduction as well as a guide to his writings.

6 Joseph Needham, "Time and knowledge in China and the West," in *The Voices of Time,* pp. 92–135.

7 Roger Penrose, "Singularities and time asymmetry," in *General Relativity,* ed. S. W. Hawking and W. Israel (Cambridge: Cambridge University Press, 1979), pp. 581, 635.

8 Eugene Wigner, *Symmetries and Reflections* (Bloomington: Indiana University Press, 1967), p. 40. See also, Joseph Needham's "Human law and the laws of nature," in *Technology, Science and Art: Common Ground* (London: Hatfield College of Technology, 1961), pp. 3–27.

9 Linguistic inertia designates the tendency of language and, by extension, of the whole human communication network including art, artifact, folkways, and mores, to resist change in the direction of the cultural process. See J. T. Fraser, "Temporal levels: sociobiological aspects of a fundamental synthesis," *Journal of Social and Biological Structures* 1 (1978): 339–55. Also, by the same author, "Out of Plato's cave: the natural history of time," *Kenyon Review,* n.s. 2 (Winter 1980): 143–62.

10 J. T. Fraser, "Toward an integrated understanding of time," *The Voices of Time,* pp. xxv–xlix.

Name Index

Subject Index

Library of Congress Cataloging in Publication Data
Fraser, J. T. (Julian Thomas), 1923–
The genesis and evolution of time.
Includes bibliographical references and index.
1. Time.   2. Relativity (Physics)   I. Title.
QB209.F7   1982   529   82-8622
ISBN 0-87023-370-X   AACR2